T0228436

Pitman Research Notes in Mathematics Series

Main Editors
H. Brezis, Université de Paris
R.G. Douglas, State University of New York at Stony Brook
A. Jeffrey, University of Newcastle upon Tyne *(Founding Editor)*

Editorial Board
R. Aris, University of Minnesota
G.I. Barenblatt, University of Cambridge
A. Bensoussan, INRIA, France
S. Bloch, University of Chicago
B. Bollobás, University of Cambridge
S. Donaldson, University of Oxford
J. Douglas Jr, Purdue University
R.J. Elliott, University of Alberta
R.P. Gilbert, University of Delaware
R. Glowinski, Université de Paris
K.P. Hadeler, Universität Tübingen
D. Jerison, Massachusetts Institute of Technology
K. Kirchgässner, Universität Stuttgart
B. Lawson, State University of New York at Stony Brook
W.F. Lucas, Claremont Graduate School
S. Mori, Kyoto University
L.E. Payne, Cornell University
G.F. Roach, University of Strathclyde
B. Simon, California Institute of Technology
S.J. Taylor, University of Virginia

Submission of proposals for consideration

Suggestions for publication, in the form of outlines and representative samples, are invited by the Editorial Board for assessment. Intending authors should approach one of the main editors or another member of the Editorial Board, citing the relevant AMS subject classifications. Alternatively, outlines may be sent directly to the publisher's offices. Refereeing is by members of the board and other mathematical authorities in the topic concerned, throughout the world.

Preparation of accepted manuscripts

On acceptance of a proposal, the publisher will supply full instructions for the preparation of manuscripts in a form suitable for direct photo-lithographic reproduction. Specially printed grid sheets can be provided and a contribution is offered by the publisher towards the cost of typing. Word processor output, subject to the publisher's approval, is also acceptable.

Illustrations should be prepared by the authors, ready for direct reproduction without further improvement. The use of hand-drawn symbols should be avoided wherever possible, in order to maintain maximum clarity of the text.

The publisher will be pleased to give any guidance necessary during the preparation of a typescript, and will be happy to answer any queries.

Important note

In order to avoid later retyping, intending authors are strongly urged not to begin final preparation of a typescript before receiving the publisher's guidelines. In this way it is hoped to preserve the uniform appearance of the series.

G P Galdi

University of Ferrara, Italy

and

J Nečas

Charles University, Czech Republic

(Editors)

Recent developments in theoretical fluid mechanics

Winter School, Paseky, 1992

CRC Press
Taylor & Francis Group
Boca Raton London New York

CRC Press is an imprint of the
Taylor & Francis Group, an **informa** business

First published 1993 by Longman Scientific & Technical

Published 2019 by CRC Press
Taylor & Francis Group
6000 Broken Sound Parkway NW, Suite 300
Boca Raton, FL 33487-2742

© 1993 by Taylor & Francis Group, LLC
CRC Press is an imprint of Taylor & Francis Group, an Informa business

First issued in paperback 2019

No claim to original U.S. Government works

ISBN 13: 978-0-367-44977-3 (pbk)
ISBN 13: 978-0-582-22684-5 (hbk)

This book contains information obtained from authentic and highly regarded sources. Reasonable efforts have been made to publish reliable data and information, but the author and publisher cannot assume responsibility for the validity of all materials or the consequences of their use. The authors and publishers have attempted to trace the copyright holders of all material reproduced in this publication and apologize to copyright holders if permission to publish in this form has not been obtained. If any copyright material has not been acknowledged please write and let us know so we may rectify in any future reprint.

Except as permitted under U.S. Copyright Law, no part of this book may be reprinted, reproduced, transmitted, or utilized in any form by any electronic, mechanical, or other means, now known or hereafter invented, including photocopying, microfilming, and recording, or in any information storage or retrieval system, without written permission from the publishers.

For permission to photocopy or use material electronically from this work, please access www. copyright.com (http://www.copyright.com/) or contact the Copyright Clearance Center, Inc. (CCC), 222 Rosewood Drive, Danvers, MA 01923, 978-750-8400. CCC is a not-for-profit organization that provides licenses and registration for a variety of users. For organizations that have been granted a photocopy license by the CCC, a separate system of payment has been arranged.

Trademark Notice: Product or corporate names may be trademarks or registered trademarks, and are used only for identification and explanation without intent to infringe.

**Visit the Taylor & Francis Web site at
http://www.taylorandfrancis.com**

**and the CRC Press Web site at
http://www.crcpress.com**

ISSN 0269-3674

British Library Cataloguing in Publication Data

A catalogue record for this book is
available from the British Library

Library of Congress Cataloging-in-Publication Data

Recent developments in theoretical fluid mechanics / G.P. Galdi and J.
  Necas, editors.
    p. cm. -- (Pitman research notes in mathematics series, ISSN
0269-3674)
    1. Fluid mechanics. I. Galdi, Giovanni P. (Giovanni Paolo),
1947-   . II. Nečas, Jindřich. III. Series.
QA901.R43 1993
532--dc20   93-4707
                                                           CIP

Contents

Foreword

The five papers collected in this volume are the content of as many series of lectures delivered by the authors at the Second Winter School in Fluid Dynamics, held in Paseky, Czech Republic, in November 29 – December 4 1992. They concern different fields in Theoretical Fluid Mechanics.

Specifically, the paper of Matsumura deals with some basic questions related to existence and stability of one–dimensional flow of a compressible fluid. The paper of Girault furnishes a detailed and comprehensive analysis of the Stokes problem in exterior domains, while that of Litvinov is dedicated to existence theory for a class of equations describing the motions of certain non–Newtonian fluids. The contribution of Rajagopal is dedicated to nonclassical fluids and to their constitutive equations as well. Finally, the paper of Galdi studies the two-dimensional exterior problem for the steady–state Navier-Stokes equations.

It is our belief that this volume will furnish new insights on classical problems as well as significant ideas on current topics.

We would like to thank all participants for their interest and for their stimulating questions and discussions during and after the lectures.

Last but not least, we acknowledge, with pleasure, a bunch of young and capable mathematicians who did most of the work to organize the School and to make it perfect from both scientific and social point of view. Specifically, we convey our sincere thanks to Drs. J. Málek (Donna Peppa), A. Novotný, M. Rokyta and M. Růžička.

April 1993

Giovanni P. Galdi
Jindřich Nečas

GIOVANNI P. GALDI

Existence and Uniqueness at Low Reynolds Number of Stationary Plane Flow of a Viscous Fluid in Exterior Domains

1. Introduction

In his celebrated paper of 1933, J. LERAY studied the existence of stationary solutions to the Navier-Stokes equations in a bounded domain Ω:

$$\Delta \mathbf{v} = \lambda \mathbf{v} \cdot \nabla \mathbf{v} + \nabla p + \mathbf{f}$$
$$\nabla \cdot \mathbf{v} = 0 \tag{1.1}$$

subject to the side condition

$$\mathbf{v} = \mathbf{v}_* \text{ at } \partial\Omega, \tag{1.2}$$

where $\partial\Omega$ is the boundary of Ω. In these (non-dimensional) relations, $\mathbf{v}$, p are velocity and pressure fields of the fluid, respectively, $\mathbf{f}$ is the body force acting on the fluid, $\mathbf{v}_*$ is a prescribed field representing the velocity of the fluid at the boundary and λ is a dimensionless parameter which can be identified with the *Reynolds number*.

As is well–known, LERAY's proof relies on an *a priori* estimate on the Dirichlet integral of $\mathbf{v}$, that is,

$$\int_\Omega \nabla \mathbf{v} : \nabla \mathbf{v} \leq M \tag{1.3}$$

where M depends only on Ω, λ and on the data $\mathbf{f}$ and $\mathbf{v}_*$. When $\mathbf{v}_* \equiv 0$ estimate (1.3) can be formally obtained by dot-multiplying $(1.1)_1$ by $\mathbf{v}$, integrating by parts over Ω and then using $(1.1)_2$[1]. In the same paper LERAY investigated also the case when Ω is an *exterior* domain, that is, the complement of a compact set. In such a situation, equations (1.1)–(1.2) must be supplemented with the following condition at infinity:

$$\lim_{|x| \to \infty} \mathbf{v}(x) = \mathbf{v}_\infty. \tag{1.4}$$

Solving a suitable sequence of problems on "invading" domains of the type $\Omega_n \equiv \Omega \cap \{|x| < R_n\}$, $n = 1, 2, \ldots$, and denoting by $\{\mathbf{v}_n\}$, $\{p_n\}$ the sequence of corresponding solutions, with the help of the uniform bound (1.3) LERAY was able to show that

---

[1] If $\mathbf{v}_* \not\equiv 0$, the derivation of (1.3) is more complicated. In particular, if $\partial\Omega$ is constituted by more than one connected component, one can show the validity of (1.3) only if the flux of $\mathbf{v}_*$ through $\partial\Omega$ satisfies some restrictions, *cf.* GALDI (1991).

1

this sequence is in fact converging to a (smooth) solution $\mathbf{v}$, p of system (1.1)–(1.2) verifying condition (1.3). Concerning condition (1.4), he proved that it is also satisfied in the case when Ω is a three-dimensional domain[2]. However, if Ω is a plane, exterior domain, the problem of proving (1.4) was left open. This can be explained in a very simple way. Actually, the only information one has on the behaviour of solutions at large distances is the property (1.3). Now, if $\Omega \subset \mathbb{R}^3$, one can show that, as a consequence of the SOBOLEV-like inequality:

$$\int_\Omega |\mathbf{v} - \mathbf{v}_\infty|^6 \leq \gamma \int_\Omega \nabla\mathbf{v} : \nabla\mathbf{v} \tag{1.5}$$

$\mathbf{v}$ must tend, in a suitable sense, to $\mathbf{v}_\infty$, cf. GALDI (1993, Chapter II, Theorem 5.1). On the other hand, if $\Omega \subset \mathbb{R}^2$, no inequality of the type (1.5) is available and, in fact, one can easily produce examples of solenoidal vector fields satisfying (1.3) and *diverging* at large spatial distances. Consider, for instance

$$\mathbf{v}(x) = (\log|x|)^\alpha \mathbf{e}_\theta, \ 0 < \alpha < 1/2 \,.$$

The problem of showing the validity of (1.4) for solutions to (1.1)-(1.2) verifying (1.3) (sometimes referred to as *D-solutions*) is a formidable one and has attracted the attention of many prominent writers, cf., e.g., GILBARG & WEINBERGER (1976, 1978), AMICK (1988). However, although these authors prove that every D-solution *does* approach (in a suitable sense) a constant vector $\mathbf{v}_0$ at large distances, they are not able to conclude $\mathbf{v}_0 = \mathbf{v}_\infty$ and, therefore, the question of proving (or disproving) the validity of (1.4) for LERAY's solutions is to be considered still open.

Let us once more emphasize that the fact that, for plane exterior flow, LERAY's solutions are not controllable at large distances, is due to the circumstance that the gradient of the velocity field is summable with exponent 2 (*cf.* (1.3)) which is *just* the dimension of the space. However, we may wonder if it is possible to determine solutions satisfying

$$\int_\Omega |\nabla\mathbf{v}|^q \leq M \text{ for some } q \in (1,2) \tag{1.6}$$

for, in such a case, by the SOBOLEV inequality we would obtain that (1.4) is satisfied (at least) in the following sense

$$\int_\Omega |\mathbf{v} - \mathbf{v}_\infty|^{2q/(2-q)} \leq \gamma_q \int_\Omega |\nabla\mathbf{v}|^q \,. \tag{1.7}$$

The aim of the present paper is to show that these solutions, in fact, exist provided the following conditions are met:

(i) $\mathbf{v}_\infty$ is non-zero;
(ii) the Reynolds number λ is sufficiently small.

---

[2] Actually, LERAY showed (1.4) for $\mathbf{v}_\infty = 0$, otherwise he only proved a weaker version of it. The proof of (1.4) in the general case is due to FINN (1959).

The main idea to prove the results, relies upon the coupling of suitable L^q-estimates obtained for a *linearized* version of (1.1), (1.2), (1.4), together with a contraction mapping argument. It is important to emphasize, however, that the procedure is not going to work unless $\mathbf{v}_\infty \neq 0$ and, even in this situation, its success is by no means evident *a priori*, as we are about to describe. To begin with, let us first take $\mathbf{v}_\infty = 0$. In such a case, the linearization associated to (1.1), (1.2), (1.4) is the Stokes one, that is,

$$\left.\begin{array}{c} \Delta\mathbf{v} = \nabla p + \mathbf{f} \\ \nabla \cdot \mathbf{v} = 0 \end{array}\right\} \ \text{in}\,\Omega \tag{1.8}$$

subject to the boundary condition (1.2) and to the condition at infinity

$$\lim_{|x|\to\infty} \mathbf{v}(x) = 0. \tag{1.9}$$

Now, it is known that an L^q theory of (1.8), (1.2), (1.9), *cf.* GALDI & SIMADER (1990), KOZONO & SOHR (1991, 1993), GALDI (1993), requires the validity of a *necessary* and *sufficient* condition on the data. Therefore, we can not use such a theory unless we show that this condition is also necessary in the nonlinear context. On the other hand, assume $\mathbf{v}_\infty \neq 0$. In view of the non-dimensional form of the problem, we may take without loss $\mathbf{v}_\infty = \mathbf{e}_1$, with $\mathbf{e}_1$ unit vector in the direction of the positive x_1–axis[3] and so, setting

$$\mathbf{u} = \mathbf{v} - \mathbf{e}_1$$

an appropriate linearized approximation for (1.1), (1.2), (1.4) is the OSEEN one, namely,

$$\left.\begin{array}{c} \Delta\mathbf{u} = \lambda\dfrac{\partial\mathbf{u}}{\partial x_1} + \nabla\pi + \mathbf{f} \\[2mm] \nabla \cdot \mathbf{u} = 0 \end{array}\right\} \ \text{in}\ \Omega \tag{1.10}$$

$$\mathbf{u} = \mathbf{v}_* - \mathbf{e}_1 \equiv \mathbf{u}_* \ \text{at}\ \partial\Omega$$

$$\lim_{|x|\to\infty} \mathbf{u}(x) = 0.$$

An L^q theory for (1.10) has been recently carried out by GALDI (1992) and it does not suffer from the drawbacks which we mentioned for the Stokes problem. As a consequence, we may use it, along with a suitable contraction mapping argument, to show existence of solutions to the nonlinear Navier-Stokes problem in the class defined by the condition (1.6), at least for small values of the Reynolds number. In doing this, however, we have to face another problem, which we are going to describe.

---

[3]If $\mathbf{v}_\infty \neq 0$, a "natural" choice of the Reynolds number is $\lambda = |\mathbf{v}_\infty|d/\nu$, where d is the diameter of the complement of Ω and ν is the coefficient of kinematical viscosity.

Specifically, we wish to find a fixed point (in a subset X of an appropriate Banach space) of the mapping

$$L : \mathbf{u} \in X \to L(\mathbf{u}) = \mathbf{v} \in X$$

where $\mathbf{v}$ solves the problem

$$\left.\begin{aligned} \Delta \mathbf{v} &= \lambda \frac{\partial \mathbf{v}}{\partial x_1} + \lambda \mathbf{u} \cdot \nabla \mathbf{u} + \nabla \pi + \mathbf{f} \\ \nabla \cdot \mathbf{v} &= 0 \end{aligned}\right\} \quad \text{in } \Omega \tag{1.11}$$

$$\mathbf{v} = \mathbf{v}_* - \mathbf{e}_1 \equiv \mathbf{u}_* \quad \text{at } \partial\Omega$$

$$\lim_{|x| \to \infty} \mathbf{v}(x) = 0.$$

In the limit of vanishing λ – as requested by the contraction argument – there will be a competition between the linear term

$$\lambda \frac{\partial \mathbf{v}}{\partial x_1} \tag{1.12}$$

and the nonlinear one

$$\lambda \mathbf{u} \cdot \nabla \mathbf{u}. \tag{1.13}$$

If, in the range of vanishing λ, the contribution of the former were negligible with respect to that of the latter, it would be very unlikely to prove existence, because the linear part in $(1.11)_1$ would then approach the Stokes system (1.8) for which, as we noticed, the procedure is not working. Fortunately, what happens is that (1.12) prevails on (1.13) and we are able to show nonlinear existence. Nevertheless, the proof of this fact is by no means trivial and relies upon the following two crucial circumstances:

(i) the validity of suitable, new *a priori* L^q-estimates for the OSEEN problem which, unlike those obtained by GALDI (1992), present an *explicit* dependence of the constant entering the estimates on the Reynolds number λ;

(ii) the component v_2 of the velocity field $\mathbf{v}$ solution to the OSEEN problem presents no "wake region" and, as a consequence, it has at large distances a "better" behaviour than that exhibited by the component v_1.

The paper is organized as follows. After recalling, in Section 2, some mathematical preliminaries and some known results on the OSEEN fundamental solution for the plane, in Section 3 we shall prove existence, uniqueness and corresponding estimates for solutions to problem (1.10) in suitable function spaces. Such results together with the contraction mapping theorem allow us, in Section 4, to show the unique resolubility of the Navier-Stokes problem (1.1), (1.2), (1.4) in the case of small Reynolds number λ. Finally, in Section 5, we shall analyze the behaviour of these solutions in the limit $\lambda \to 0$ and shall prove that they tend to solutions of the

4

corresponding Stokes system (1.8), (1.2). Concerning the behaviour at infinity, we shall show that the limit solution to the Stokes system verifies the same condition of the solution to the Navier-Stokes problem *if and only if* the data satisfy a certain compatibility condition. In the case $\mathbf{f} = 0$, $\mathbf{v}_* \neq 0$ and Ω the exterior of a unit circle, this condition becomes

$$\int_{\partial\Omega} (\mathbf{v}_* - \mathbf{v}_\infty) = 0.$$

Last but not least, we wish to recall that, by means of a completely different approach based on the estimates of the GREEN's tensor-solution, FINN & SMITH (1967) have given the first (and, so far, the only) existence result for problem (1.1), (1.2), (1.4) with $\mathbf{v}_\infty \neq 0$ and for small λ. Under more stringent regularity assumptions on $\mathbf{v}_*$ and Ω than those made in the present paper and taking $\mathbf{f} \equiv 0$, these authors prove existence in the class of *physically reasonable* solutions for which the velocity field $\mathbf{v}$ behaves, at large distances, as the OSEEN fundamental solution. In particular, $\mathbf{v}$ obeys the uniform estimate

$$\mathbf{v}(x) = O(|x|^{-1/2}), \quad \text{as } |x| \to \infty.$$

In these regards, it is worth emphasizing that the existence theory developed in the present paper contains that of FINN & SMITH as a special case since, if $\mathbf{f}$ is of bounded support, our solutions are also physically reasonable, as a consequence of the recent result of AMICK (1991)[4] . However, because of the lack of an appropriate uniqueness theory (*cf.* Remark 4.2 and FINN & SMITH (1967, Section 7)) we are not able to prove if our solutions coincide with those of FINN & SMITH (for small λ) and we leave it as an interesting open question.

2. Preliminaries

Let us first introduce some notations. $\mathbb{N}$ is the set of all positive integers. $\mathbb{R}$ is the real line and $\mathbb{R}^2$ is the two–dimensional Euclidean space. The canonical basis in $\mathbb{R}^2$ is indicated by $\{\mathbf{e}_1, \mathbf{e}_2\}$. The disc of radius R centered at the origin will be denoted by B_R. By Ω we shall always denote a domain (open connected set) in $\mathbb{R}^2$. By $\bar{\Omega}$ we shall mean the closure of Ω and by $\partial\Omega$ its boundary. We also set $\Omega^c = \mathbb{R} - \Omega$. For $\mathcal{B} \subset \mathbb{R}^2$ we indicate by $\delta(\mathcal{B})$ its diameter. If Ω is a domain which is the complement of a (non-necessarily connected), compact set Ω^c with *bounded* boundary (*i.e.*, Ω is an *exterior* domain), taking the origin of coordinates into the interior of Ω^c, for $R > \delta(\Omega^c)$ and $R_2 > R_1 > \delta(\Omega^c)$ we put

$$\Omega_R = \{x \in \Omega : |x| < R\},$$
$$\Omega^R = \{x \in \Omega : |x| > R\},$$
$$\Omega_{R_1, R_2} = \{x \in \Omega : R_2 > |x| > R_1\}.$$

---

[4] The fact that solutions determined in this paper are physically reasonable can be proved independently of AMICK's result, but we shall not include this here.

We indicate by $C_0^\infty(\Omega)$ the class of functions in Ω wich are indefinitely differentiable and of compact support in Ω[5]. For $k = 1, 2$ we set $D_k = \partial/\partial x_k$. Likewise, for $\alpha = (\alpha_1, \alpha_2)$, $\alpha_i \geq 0$, we let

$$D^\alpha = \frac{\partial^{\alpha_1 + \alpha_2}}{\partial x_1^{\alpha_1} \partial x_2^{\alpha_2}}, \ |\alpha| = \alpha_1 + \alpha_2.$$

By $W^{m,q}(\Omega)$, $m \in \mathbb{N} \cup \{0\}$, $q \in [1, \infty]$, we shall indicate the SOBOLEV space of order (m, q) endowed with the norm

$$\|u\|_{m,q,\Omega} = \left(\sum_{|\alpha|=0}^m \int_\Omega |D^\alpha u|^q \right)^{1/q},$$

where the subscript Ω will be omitted if no confusion arises[6]. We have $W^{0,q}(\Omega) = L^q(\Omega)$ and set $\|u\|_{0,q,\Omega} \equiv \|u\|_{q,\Omega}$. The duality pairing in $L^q(\Omega)$ is denoted by $(\cdot, \cdot)$, that is

$$(u, v) = \int uv \, dx, \quad u \in L^q(\Omega), \ v \in L^{q'}(\Omega), \ 1/q' = 1 - 1/q, .[7]$$

The completion of $C_0^\infty(\Omega)$ in the norm of $W^{m,q}(\Omega)$ is denoted by $W_0^{m,q}(\Omega)$ and the dual space of $W_0^{m,q}(\Omega)$ is indicated by $W^{-m,q'}(\Omega)$, $q' = q/(q-1)$. Finally, by $W^{m-1/q,q}(\partial\Omega)$ we mean the trace space at $\partial\Omega$ of functions from $W^{m,q}(\Omega)$ and indicate by $\|\cdot\|_{m-1/q,q(\partial\Omega)}$ the associated norm, *cf.* NEČAS (1967, Chapitre 2, §5).

For $m \in \mathbb{N} \cup \{0\}$ and $q \in (1, \infty)$ we define the *homogeneous* SOBOLEV space

$$D^{m,q} = D^{m,q}(\Omega) = \left\{ u \in L^1_{loc}(\Omega) : \ D^\ell u \in L^q(\Omega), |\ell| = m \right\}.$$

One can prove that if $u \in D^{m,q}(\Omega)$ then

$$D^\ell u \in L^q(\Omega'), 0 \leq |\ell| \leq m, \quad \text{for all compact } \Omega' \text{ with } \bar{\Omega}' \subset \Omega,$$

or, in a shorter notation,

$$u \in W_{loc}^{m,q}(\Omega).$$

---

[5] As a rule, if Y denotes a space of scalar functions, we shall use the same symbol to denote the space of vector functions with components in Y.

[6] Unless their use clarifies the context, we shall omit also the infinitesimal volume and surface elements in the integrals.

[7] We shall use the same notation for vectors $\mathbf{u}$, $\mathbf{v}$, namely, we set

$$(\mathbf{u}, \mathbf{v}) = \sum_{i=1}^2 \int_\Omega u_i v_i.$$

If, in addition, Ω is locally lipschitzian, we also have $D^\ell u \in L^q(\Omega')$, $0 \le |\ell| \le m$, for all bounded domains $\Omega' \subset \Omega$ or, in a shorter notation,

$$u \in W^{m,q}_{loc}(\bar{\Omega}).$$

Thus, if Ω is bounded and locally lipschitzian, $D^{m,q}(\Omega)$ is algebraically isomorphic to $W^{m,q}(\Omega)$. In $D^{m,q}(\Omega)$ we introduce the norm

$$|u|_{m,q,\Omega} \equiv \left(\sum_{|\alpha|=m} \int_\Omega |D^\alpha u|^q \right)^{1/q},$$

where, as before, the subscript Ω will be omitted if no confusion arises. It is simple to show that $\{D^{m,q}, |\cdot|_{m,q}\}$ is a complete normed space, provided we identify two functions $u_1, u_2 \in D^{m,q}(\Omega)$ whenever $|u_1 - u_2|_{m,q} = 0$, that is, u_1 and u_2 differ, at most, by a polynomial of degree $m - 1$.

We also define the spaces $D_0^{m,q}(\Omega)$ as the completion of $C_0^\infty(\Omega)$ in the seminorm $|\cdot|_{m,q}$.

For Ω an exterior domain, one can characterize the behaviour at large distances of functions from $D^{m,q}$. In the sequel, we shall need the following special result whose proof can be found in GALDI (1993, Chapter II, Theorem 5.1).

Lemma 2.1. *Let Ω be an exterior domain with a locally lipschitzian boundary. Then, for any $u \in D^{1,q}(\Omega)$, $1 \le q < 2$, there exists a uniquely determined $u_0 \in \mathbb{R}$ such that*

$$\lim_{|x|\to\infty} \int_0^{2\pi} |u(|x|, \theta) - u_0| \, d\theta = 0.$$

Moreover,

$$u - u_0 \in L^{2q/(2-q)}(\Omega)$$

and the following inequality holds

$$\|u - u_0\|_{2q/(2-q)} \le \gamma |u|_{1,q}$$

with a constant $\gamma = \gamma(\Omega, q)$.

The next lemma is a particular case of a result due to GALDI & SIMADER (1990, Theorem 1.1).

Lemma 2.2. *Let Ω satisfy the assumption of Lemma 2.1 and let $u \in D^{1,q}(\Omega)$, $1 < q < 2$. Then $u \in D_0^{1,q}(\Omega)$ if and only if u has zero trace at $\partial\Omega$ and $u_0 = 0$, with u_0 defined in Lemma 2.1.*

Finally, we introduce a space of vector functions which will play a fundamental role in the existence theory we shall subsequently develop. For $q \in (1, 6/5]$, we indicate by $\mathcal{C}_q$ the class of those vector functions $\mathbf{u} = (u_1, u_2)$ defined in Ω such that:

$$u_2 \in L^{2q/(2-q)}(\Omega) \cap D^{1,q}(\Omega)$$
$$\mathbf{u} \in L^{3q/(3-2q)}(\Omega) \cap D^{2,q}(\Omega).$$

For $\mathbf{u} \in \mathcal{C}_q$ and $\lambda > 0$, we put

$$\langle \mathbf{u} \rangle_{\lambda,q} \equiv \lambda(\|u_2\|_{2q/(2-q)} + |u_2|_{1,q}) + \lambda^{2/3}\|\mathbf{u}\|_{3q/(3-2q)} + \lambda^{1/3}|\mathbf{u}|_{1,3q/(3-q)} \qquad (2.1)$$

and, as usual, we write $\langle \mathbf{u} \rangle_{\lambda,q,\Omega}$ when we need to specify the domain where the norm (2.1) is defined.

Remark 2.1. For $\mathbf{u} \in \mathcal{C}_q$, it holds

$$\lim_{|x|\to\infty} \mathbf{u}(x) = 0, \text{ uniformly.}$$

Actually, by Lemma 2.1, $\mathbf{u} \in D^{1,2q/(2-q)}(\Omega)$. Being $2q/(2-q) > 2$ and since $\mathbf{u} \in L^{3q/(3-2q)}(\Omega)$, the property follows from known results, $cf.$, $e.g.$, GALDI (1993, Chapter II, Theorem 5.1). If $\mathbf{u}$ has only finite norm (2.1), then

$$\lim_{|x|\to\infty} \int_0^{2\pi} |\mathbf{u}(|x|,\theta)| d\theta = 0. \qquad (2.2)$$

In fact, if $q \in (1, 6/5)$ then $\mathbf{u} \in D^{1,r}(\Omega)$ for some $r < 2$ and so, from Lemma 2.1 and from the condition $\mathbf{u} \in L^{3q/(3-2q)}(\Omega)$, we find (2.2). If $q = 6/5$ it is $\mathbf{u} \in D^{1,2}(\Omega) \cap L^6(\Omega)$ and we proceed as follows. Set

$$f(r) = \int_0^{2\pi} |\mathbf{u}(r,\theta)|^2 d\theta, \; r = |x|.$$

Then there exists a sequence $\{r_n\} \subset \mathbb{N}$ accumulating at infinity such that

$$\lim_{n\to\infty} f(r_n) = 0.$$

For all sufficiently large r we may find $n \in \mathbb{N}$ such that

$$f(r) = f(r_n) + 2\int_{r_n}^r \int_0^{2\pi} \mathbf{u}(\rho,\theta) \cdot \frac{\partial \mathbf{u}(\rho,\theta)}{\partial \rho} d\rho d\theta.$$

Applying HÖLDER inequality at the left hand side of this relation it follows

$$|f(r)| \leq |f(r_n)| + 2\|\mathbf{u}\|_{6,\Omega_{r_n}} |\mathbf{u}|_{1,2,\Omega_{r_n}} \left(\int_{r_n}^\infty \rho^{-2} d\rho\right)^{1/3}$$

which again proves (2.2).

In the final part of this section, we recall some basic and well-known properties of the OSEEN fundamental tensor. For more detailed information and for the proof of the stated results, we refer the reader to GALDI (1993, Chapter VII, Section 3). Following OSEEN (1927, §4), for $x, y \in \mathbb{R}^2$ we denote by $\mathbf{E}$, $\mathbf{e}$ tensor and vector fields, respectively, such that $(i, j = 1, 2)$

$$E_{ij}(x - y) = \left(\delta_{ij} \Delta - \frac{\partial^2}{\partial y_i \partial y_j} \right) \Phi(x - y)$$

$$e_{ij}(x - y) = -\frac{\partial}{\partial y_j} \left(\Delta - \lambda \frac{\partial}{\partial y_1} \right) \Phi(x - y) \tag{2.3}$$

where

$$\Phi(x - y) = -\frac{1}{2\pi\lambda} \int_{x_1}^{y_1} \left\{ \log \sqrt{(\tau - x_1)^2 + (x_2 - y_2)^2} \right.$$

$$\left. + K_0 \left(\frac{\lambda}{2} \sqrt{(\tau - x_1)^2 + (x_2 - y_2)^2} \right) e^{-\lambda(\tau - x_1)} \right\} d\tau, \tag{2.4}$$

and $K_0(z)$ is the modified Bessel function of the second kind of order zero. Moreover,

$$e_j(x - y) = \frac{1}{2\pi} \frac{x_j - y_j}{|x - y|^2}, \quad j = 1, 2. \tag{2.5}$$

By a direct (and tedious) calculation, one can show that for all $x \neq y$[8]

$$\left(\Delta + \lambda \frac{\partial}{\partial y_1} \right) E_{ij}(x - y) - \frac{\partial}{\partial y_i} e_j(x - y) = 0$$

$$\frac{\partial}{\partial y_l} E_{lj}(x - y) = 0,$$

while, as $|x - y| \to 0$, $\mathbf{E}(x - y)$ becomes singular with the same order as the Stokes fundamental tensor $\mathbf{U}$ defined, this latter, as

$$U_{ij}(x - y) = -\frac{1}{4\pi} \left[\delta_{ij} \log \left(\frac{1}{|x - y|} \right) + \frac{(x_i - y_i)(x_j - y_j)}{|x - y|^2} \right], \quad i, j = 1, 2. \tag{2.6}$$

Specifically, one can prove the following asymptotic relation,

$$E_{ij}(x - y) = U_{ij}(x - y) - \frac{1}{4\pi} \delta_{ij} \log \frac{1}{\lambda} + o(1), \quad \text{as} \quad \lambda|x - y| \to 0. \tag{2.7}$$

---

[8] We adopt the Einstein summation convention over repeated indices.

We shall now collect some properties of $\mathbf{E}$ which will be frequently used throughout the paper. Denote by $\mathbf{E}(x - y; \lambda)$ the OSEEN tensor corresponding to the Reynolds number λ. We have the *homogeneity property*

$$\mathbf{E}(x - y; \lambda) = \mathbf{E}(\lambda(x - y); 1) \tag{2.8}$$

Set

$$\begin{aligned} \mathbf{E}_1 &= (E_{11}, E_{12}) \\ \mathbf{E}_2 &= (E_{12}, E_{22}). \end{aligned} \tag{2.9}$$

Denoting by $\mathcal{A}$ the exterior of any circle centered at the origin, we have the following uniform bounds[9]

$$D_l \mathbf{E}_i(x) \leq \left\{ \begin{array}{ll} c|x|^{-1} & \text{if i=1} \\ c|x|^{-3/2} & \text{if i=2} \end{array} \right. \quad x \in \mathcal{A}, \ l = 1, 2, \tag{2.10}$$

and

$$D_k D_l \mathbf{E}_i(x) \leq \left\{ \begin{array}{ll} c|x|^{-3/2} & \text{if i=1} \\ c|x|^{-2} & \text{if i=2} \end{array} \right. \quad x \in \mathcal{A}, \ \ l, k = 1, 2. \tag{2.11}$$

In addition, the following summability properties hold

$$\begin{aligned} \mathbf{E}_1 &\in L^q(\mathcal{A}), \quad \text{for all } q > 3 \\ \mathbf{E}_2 &\in L^q(\mathcal{A}), \quad \text{for all } q > 2, \\ \frac{\partial \mathbf{E}_i}{\partial x_1} &\in L^q(\mathcal{A}) \quad \text{for all } \ q > 1, \ i = 1, 2 \\ \frac{\partial \mathbf{E}_i}{\partial x_2} &\in L^q(\mathcal{A}) \quad \text{for all } q > 3/2, \ i = 1, 2 \\ D_k D_l \mathbf{E}_i &\in L^q(\mathcal{A}) \quad \text{for all } q > 1 q, \ i, k, l = 1, 2. \end{aligned} \tag{2.12}$$

3. Existence, Uniqueness and L^q-Estimates for the Oseen Problem

The objective of this section is to prove existence, uniqueness and the validity of corresponding estimates in appropriate homogeneous SOBOLEV spaces for solutions to problem (1.10) under different assumptions on the data. First of all, we begin to collect some known results which are particular cases of those shown by GALDI (1992) and GALDI (1993, Chapter VII, Sections 4, 5 and 7).

---

[9] Notice that $\mathbf{E}$ depends on x, y only through $x - y$.

Lemma 3.1. (GALDI *(1992, Theorem 2.1)). Consider the Oseen problem*

$$\left.\begin{aligned} \Delta \mathbf{v} = \lambda \frac{\partial \mathbf{v}}{\partial x_1} + \nabla p + \mathbf{f} \\ \nabla \cdot \mathbf{v} = 0 \end{aligned}\right\} \quad in \ \mathbb{R}^2 \tag{3.1}$$

$$\lim_{|x| \to \infty} \mathbf{v} = 0.$$

Then, if

$$\mathbf{f} \in L^q(\mathbb{R}^2), \quad q \in (1, 6/5],$$

there exists one and only one solution $\mathbf{v}$, π *to (3.1) such that*

$$\mathbf{v} \in \mathcal{C}_q, \quad p \in D^{1,q}(\mathbb{R}^2).$$

This solution satisfies the estimate

$$\langle \mathbf{v} \rangle_{\lambda,q} + |\mathbf{v}|_{2,q} + |\pi|_{1,q} \leq c \|\mathbf{f}\|_q$$

where $c = c(q)$, *for all* $\lambda \in (0, 1]$, *and* $\langle \cdot \rangle_{\lambda,q}$ *is defined in (2.1).*

We recall that the *stress tensor* $\mathbf{T} = \mathbf{T}(\mathbf{v}, p)$ associated to a motion $\mathbf{v}$, p is defined by

$$T_{ij} = \frac{1}{2} \left(\frac{\partial v_i}{\partial x_j} + \frac{\partial v_j}{\partial x_i} \right) - p \delta_{ij}, \quad i, j = 1, 2.$$

The *total force* exerted by the fluid on the body Ω^c is then given by

$$\mathcal{T}(\mathbf{v}) = - \int_{\partial \Omega} \mathbf{T}(\mathbf{v}, p) \cdot \mathbf{n} \tag{3.2}$$

with $\mathbf{n}$ unit outer normal to $\partial \Omega$. We have

Lemma 3.2. (GALDI *(1993, Chapter VII, Theorems 5.1, 6.2 and 8.1)). Let* Ω *be an exterior domain of class* C^2. *Consider the Oseen problem*

$$\left.\begin{aligned} \Delta \mathbf{u} = \lambda \frac{\partial \mathbf{u}}{\partial x_1} + \nabla \pi \\ \nabla \cdot \mathbf{u} = 0 \end{aligned}\right\} \quad in \ \Omega \tag{3.3}$$

$$\mathbf{u} = \mathbf{u}_* \quad at \ \partial \Omega$$

$$\lim_{|x| \to \infty} \mathbf{u}(x) = 0.$$

Then, given

$$\mathbf{u}_* \in W^{2-1/q,q}(\partial \Omega), \quad q \in (1, 6/5],$$

11

there exists one and only one solution $\mathbf{u}$, π to (3.3) such that

$$\mathbf{u} \in \mathcal{C}_q, \quad \pi \in D^{1,q}(\Omega).$$

Also, $\mathbf{u}$ admits the following representation for all $x \in \Omega$

$$u_j(x) = \int_{\partial\Omega} \mathbf{u}_*(z) \cdot \mathbf{T}[\mathbf{E}_j(x - z; \lambda), e_j(x - z)] \cdot \mathbf{n}(z)d\sigma_z$$

$$+ \lambda \int_{\partial\Omega} \mathbf{u}_*(z) \cdot \mathbf{E}_j(x - z; \lambda)n_1(z)d\sigma_z$$

$$+ \int_{\partial\Omega} [\mathbf{E}_j(x - z; \lambda) \cdot \mathbf{T}(\mathbf{u}, \pi) \cdot \mathbf{n}(z)d\sigma_z.$$

Moreover, for all $R > \delta(\Omega^c)$ the solution satisfies

$$\langle\mathbf{u}\rangle_{\lambda,q} + \|\pi\|_{2,\Omega_R} + |\mathbf{u}|_{2,q} + |\pi|_{1,q} \le c_1\|\mathbf{u}_*\|_{2-1/q,q(\partial\Omega)}$$

where $c_1 = c_1(\Omega, q, R)$, for all $\lambda \in (0,1]$. Finally, there is $\lambda_1 > 0$ and $c_2 = c_2(\Omega, R, q, \lambda_1)$ such that for all $\lambda \in (0, \lambda_1]$

$$|\mathcal{T}(\mathbf{u})| \le c_2|\log \lambda|^{-1}\|\mathbf{u}_*\|_{2-1/q,q(\partial\Omega)}.$$

We are now in a position to prove the next result which furnishes a sharper estimate for solutions to problem (3.3).

Lemma 3.3. *Let the assumptions of Lemma 3.2 be satisfied. Then, there exists a $\lambda_0 > 0$ such that for all $\lambda \in (0, \lambda_0]$ the solution $\mathbf{u}$ obeys the estimate*

$$\langle\mathbf{u}\rangle_{\lambda,q} \le c\lambda^{2(1-1/q)}|\log \lambda|^{-1}\|\mathbf{u}_*\|_{2-1/q,q(\partial\Omega)} \tag{3.4}$$

with $c = c(\Omega, q, \lambda_0)$.

PROOF : First of all we notice that, from Lemma 3.2, it follows, in particular,

$$\|\mathbf{u}\|_{2,q,\Omega_R} + \|\pi\|_{1,q,\Omega_R} \le c_1\|\mathbf{u}_*\|_{2-1/q,q(\partial\Omega)} \tag{3.5}$$

with c_1 independent of $\lambda \in (0,1]$. Without loss of generality, we assume $\Omega^c \subset B_{1/2}$, with the origin of coordinates taken in the interior of Ω^c. From the well-known SOBOLEV embedding theorems, *cf.* NEČAS (1967, Chapitre 2, §3), it is

$$\|u_2\|_{2q/(2-q),\Omega_1} + |u_2|_{1,q,\Omega_1} + \|\mathbf{u}\|_{3q/(3-2q),\Omega_1} + |\mathbf{u}|_{1,3q/(3-q),\Omega_1} \le c_2\|\mathbf{u}\|_{2,q,\Omega_1} \tag{3.6}$$

and so, since $2(1 - 1/q) < 1/3$ for $q \in (1, 6/5)$, from (3.5), (3.6) and recalling (2.1) we find for some c_3 independent of $\lambda \in (0,1]$

$$\langle\mathbf{u}\rangle_{\lambda,q,\Omega_1} \le c_3\lambda^{2(1-1/q)+\epsilon}\|\mathbf{u}_*\|_{2-1/q,q(\partial\Omega)} \tag{3.7}$$

where $\epsilon = 1/3 - 2(1 - 1/q) > 0$. From the representation of Lemma 3.2 we deduce for $j = 1, 2$

$$
\begin{aligned}
u_j(x) = &-\mathcal{T}(\mathbf{u}) \cdot \mathbf{E}_j(x; \lambda) \\
&+ \int_{\partial\Omega} \mathbf{u}_*(z) \cdot \mathbf{T}[\mathbf{E}_j(x - z; \lambda), e_j(x - z)] \cdot \mathbf{n}(z) d\sigma_z \\
&+ \lambda \int_{\partial\Omega} \mathbf{u}_*(z) \cdot \mathbf{E}_j(x - z; \lambda) n_1(z) d\sigma_z \\
&+ \int_{\partial\Omega} [\mathbf{E}_j(x - z; \lambda) - \mathbf{E}_j(x; \lambda)] \cdot \mathbf{T}(\mathbf{u}, \pi) \cdot \mathbf{n}(z) d\sigma_z
\end{aligned}
\tag{3.8}
$$

with $\mathcal{T}(\mathbf{u})$ given in (3.2) and $\mathbf{E}_i$ defined in (2.9) Recalling that $\Omega^c \subset B_{1/2}$, from (3.8) and from the mean-value theorem applied to $D^\alpha(\mathbf{E}_j(x - z; \lambda) - \mathbf{E}_j(x; \lambda))$, $\alpha = 0, 1$, we derive for all $x \in \Omega^1$

$$
\begin{aligned}
|u_j(x)| \leq &|\mathcal{T}(\mathbf{u})||\mathbf{E}_j(x; \lambda)| + D\{\lambda \sup_{z \in \Omega_{1/2}} |\mathbf{E}_j(x - z; \lambda)| \\
&+ \sup_{z \in \Omega_{1/2}} [|e_j(x - z)| + |\nabla_x \mathbf{E}_j(x - z; \lambda)|]\} \\
|\nabla \mathbf{u}_j(x)| \leq &|\mathcal{T}(\mathbf{u})||\nabla \mathbf{E}_j(x; \lambda)| + D\{\lambda \sup_{z \in \Omega_{1/2}} |\nabla_x \mathbf{E}_j(x - z; \lambda)| \\
&+ \sup_{z \in \Omega_{1/2}} [|\nabla_x e_j(x - z)| + |D_x^2 \mathbf{E}_j(x - z; \lambda)|]\}
\end{aligned}
\tag{3.9}
$$

where

$$
D = \|\nabla \mathbf{u}\|_{1,\partial\Omega} + \|\pi\|_{1,\partial\Omega} + \|\mathbf{u}_*\|_{1,\partial\Omega}.
\tag{3.10}
$$

Taking into account (2.8), from $(3.9)_1$ we find, with $y = \lambda x$,

$$
\begin{aligned}
|u_j(x)| \leq &|\mathcal{T}(\mathbf{u})||\mathbf{E}_j(y; 1)| + \lambda D\{\sup_{z \in \Omega_{1/2}} [|\mathbf{E}_j(y - \lambda z; 1)| \\
&|e_j(y - \lambda z)| + |\nabla_y \mathbf{E}_j(y - \lambda z; 1)|]\} \\
\leq &|\mathcal{T}(\mathbf{u})||\mathbf{E}_j(y; 1)|\lambda D\{\sup_{|z| \leq \lambda/2} [|\mathbf{E}_j(y - z; 1)| \\
&|e_j(y - z)| + |\nabla_y \mathbf{E}_j(y - z; 1)|]\}
\end{aligned}
$$

and so

$$
\begin{aligned}
\|u_j\|_{t,\Omega^1}^t \leq 2^t \lambda^{-2} \Big\{ &|\mathcal{T}(\mathbf{u})|^t \|\mathbf{E}_j(y; 1)\|_{t,\mathbb{R}^2}^t \\
&+ \lambda^t D^t \int_{|y| \geq \lambda} \{\sup_{|z| \leq \lambda/2} [|\mathbf{E}_j(y - z; 1)| \\
&+ |e_j(y - z)| + |\nabla_y \mathbf{E}_j(y - z; 1)|]\}^t dy \Big\} .
\end{aligned}
\tag{3.11}
$$

To estimate the integral at the right hand side of (3.11) we observe that

$$\int_{|y|\geq\lambda}\sup_{|z|\leq\lambda/2}|\mathbf{E}_j(y-z;1)|^t\,dy \leq \int_{2\geq|y|\geq\lambda}\sup_{|z|\leq\lambda/2}|\mathbf{E}_j(y-z;1)|^t\,dy$$

$$+\int_{|y|\geq 2}\sup_{|z|\leq 1/2}|\mathbf{E}_j(y-z;1)|^t\,dy. \tag{3.12}$$

From (2.6)-(2.7) we have

$$|\mathbf{E}(y-z;1)| \leq c(|\log|y-z||+1), \quad |y|,|z| \leq 2$$

and since

$$|z| \leq \lambda/2, |y| \geq \lambda \text{ implies } |y-z| \geq |y|, \tag{3.13}$$

it follows

$$\int_{2\geq|y|\geq\lambda}\sup_{|z|\leq\lambda/2}|\mathbf{E}_j(y-z;1)|^t\,dy \leq c_4 \tag{3.14}$$

with c_4 independent of $\lambda \in (0,1]$. Furthermore, since by the mean value theorem it is

$$|(E_{ij}(y-z)-E_{ij}(y))| = |z_l\frac{\partial E_{ij}(y-\beta z)}{\partial x_l}|, \quad \beta \in (0,1),$$

from (2.10) it follows for all sufficiently large $|y|$ ($\geq |y_0|$, say)

$$\int_{|y|\geq|y_0|}\sup_{|z|\leq 1/2}|\mathbf{E}_j(y-z;1)|^t\,dy \leq c\int_{|y|\geq|y_0|}(|\mathbf{E}_j(y;1)|^t + |y|^{-tr_j})\,dy.$$

where

$$r_j = \begin{cases} 1 & \text{if } j=1 \\ 3/2 & \text{if } j=2. \end{cases}$$

From the local regularity of $\mathbf{E}$ we then conclude

$$\int_{|y|\geq 2}\sup_{|z|\leq 1/2}|\mathbf{E}_j(y-z;1)|^t\,dy \leq c_5 \tag{3.15}$$

with c_5 independent of $\lambda \in (0,1]$, for all values of t such that

$$\mathbf{E}_j(y;1) \in L^t(|y| \geq |y_0|). \tag{3.16}$$

Collecting (3.12), (3.14) and (3.15) it follows

$$\int_{|y|\geq\lambda}\sup_{|z|\leq\lambda/2}|\mathbf{E}_j(y-z;1)|^t\,dy \leq c_6 \tag{3.17}$$

14

with c_6 independent of $\lambda \in (0,1]$ and for all values of t for which (3.16) holds. In addition, from (2.5) and (3.13) we infer for all $t > 2$

$$\int_{|y|\geq\lambda} \sup_{|z|\leq\lambda/2} |\mathbf{e}_j(y-z;1)|^t dy \leq c_7 \left\{ \int_{2\geq|y|\geq\lambda} |y|^{-t}\, dy \right.$$
$$\left. + \int_{|y|\geq 2} \sup_{|z|\leq 1/2} |y-z|^{-t} dy \right\} \tag{3.18}$$
$$\leq c_8(\lambda^{2-t}+1).$$

Likewise,

$$\int_{|y|\geq\lambda} \sup_{|z|\leq\lambda/2} |\nabla_y\mathbf{E}_j(y-z;1)|^t\, dy \leq \int_{2\geq|y|\geq\lambda} \sup_{|z|\leq\lambda/2} |\nabla_y\mathbf{E}_j(y-z;1)|^t\, dy$$
$$+ \int_{|y|\geq 2} \sup_{|z|\leq 1/2} |\nabla_y\mathbf{E}_j(y-z;1)|^t\, dy. \tag{3.19}$$

From (2.8) we have

$$|\nabla\mathbf{E}(y-z;1)| \leq c_9|y-z|^{-1}, \quad |y|,|z| \leq 2$$

and so, by (3.13), it follows

$$\int_{2\geq|y|\geq\lambda} \sup_{|z|\leq\lambda/2} |\nabla_y\mathbf{E}_j(y-z;1)|^t dy \leq c_{10} \int_{2\geq|y|\geq\lambda} |y|^{-t} dy \leq c_{11}(1+\lambda^{2-t}). \tag{3.20}$$

Also, employing the asymptotic properties (2.11) of the second derivatives we are able to establish, exactly as we did for (3.15), the following estimate:

$$\int_{|y|\geq 2} \sup_{|z|\leq 1/2} |\nabla_y\mathbf{E}_j(x-z;1)|^t dy \leq c_{12} \tag{3.21}$$

for those values of t such that

$$\nabla\mathbf{E}_j(y;1) \in L^t(|y|\geq|y_0|). \tag{3.22}$$

Thus, from (3.19)–(3.21) we recover

$$\int_{|y|\geq\lambda} \sup_{|z|\leq\lambda/2} |\nabla_y\mathbf{E}_j(y-z;1)|^t dy \leq c_{13}(1+\lambda^{2-t}), \tag{3.23}$$

with c_{13} independent of λ and with t satisfying (3.22). From (3.11), (3.17), (3.18) and (3.23) we find

$$\|u_j\|_{t,\Omega^1} \leq c_{14}\left(\lambda^{-2/t}|\mathcal{T}(\mathbf{u}))| + (1+\lambda^{1-2/t})D\right),$$

for all $\lambda \in (0, 1]$, for all t for which (3.16) and (3.22) hold and with a constant c_{14} independent of λ. Being $t > 2$, from this latter inequality we deduce

$$\|u_j\|_{t,\Omega^1} \leq 2c_{14} \left(\lambda^{-2/t} |\mathcal{T}(\mathbf{u}))| + D \right). \tag{3.24}$$

Estimate (3.24) furnishes, in particular,

$$\lambda \|u_2\|_{2q/(2-q),\Omega^1} + \lambda^{2/3} \|\mathbf{u}\|_{3q/(3-2q),\Omega^1} \leq c \left(\lambda^{2(1-1/q)} |\mathcal{T}(\mathbf{u}))| + \lambda^{2/3} D \right), \tag{3.25}$$

with c independent of $\lambda \in (0, 1]$[10]. In a completely analogous way, starting with (3.9$_2$) we show

$$|u_j|^\tau_{1,\tau,\Omega^1} \leq 2^\tau \lambda^{\tau-2} \left\{ |\mathcal{T}(\mathbf{u})|^\tau \|\nabla \mathbf{E}_j(y; 1)\|^\tau_{\tau,\mathbb{R}^2} \right.$$
$$+ \lambda^\tau D^\tau \int_{|y| \geq \lambda} \left\{ \sup_{|z| \leq \lambda/2} [|\nabla_y \mathbf{E}_j(y - z; 1)| \right.$$
$$\left. \left. + |\nabla_y \mathbf{e}_j(y - z)| + |D^2_y \mathbf{E}_j(y - z; 1)|] \right\}^\tau dy \right\}.$$

Therefore, noting that, by (2.5) it is

$$|\nabla_y \mathbf{e}_j(y - z)| \leq c_{15} |y - z|^{-2}, \tag{3.26}$$

and that, by (2.11),

$$\int_{2 \geq |y| \geq \lambda} \sup_{|z| \leq \lambda/2} |D^2_y \mathbf{E}_j(y - z; 1)|^\tau dy \leq c_{16} \int_{2 \geq |y| \geq \lambda} |y|^{-2\tau} dy \leq c_{17}(1 + \lambda^{2(1-\tau)}),$$

using the same procedure employed to obtain (3.24) we arrive at

$$|u_j|_{1,\tau,\Omega^1} \leq c_{18} \left(\lambda^{1-2/\tau} |\mathcal{T}(\mathbf{u})| + D \right),$$

for all values of $\lambda \in (0, 1]$ and for all values of τ such that

$$\nabla \mathbf{E}_j(y; 1) \in L^\tau(|y| \geq |y_0|),$$
$$D^2 \mathbf{E}_j \in L^\tau(|y| \geq |y_0|).$$

Thus, from (2.12), (3.26) and observing that $\tau > 1$ we derive, in particular,

$$\lambda |u_2|_{1,q,\Omega^1} + \lambda^{1/3} |\mathbf{u}|_{1,3q/(3-q),\Omega^1} \leq c \left(\lambda^{2(1-1/q)} |\mathcal{T}(\mathbf{u})| + \lambda^{1/3} D \right), \tag{3.27}$$

---

[10]Observe that $\lambda \leq \lambda^{2/3}$ for $0 \leq \lambda \leq 1$.

for some $c = c(\Omega)$ and $q \in (1, 6/5)$. The next step is to increase D (defined in (3.10)) in terms of the boundary norm $\|\mathbf{u}_*\|_{2-1/q,q(\partial\Omega)}$. Using GAGLIARDO's trace theorem, cf. NEČAS (1967, Chapitre 2, §5), we have

$$D \leq c_{19} \left(\|\mathbf{u}\|_{2,q,\Omega_1} + \|\pi\|_{1,q,\Omega_1} + \|\mathbf{u}_*\|_{2-1/q,q(\partial\Omega)} \right)$$

and so this inequality together with (3.5) implies

$$D \leq c_{20} \|\mathbf{u}_*\|_{2-1/q,q(\partial\Omega)}.$$

Noticing that $1/3 > 2(1 - 1/q)$ for $q \in (1, 6/5)$, from (3.7), (3.25) and (3.27) we obtain

$$\langle \mathbf{u} \rangle_{\lambda,q} \leq c_{21} \lambda^{2(1-1/q)} \left(|\mathcal{T}(\mathbf{u})| + \lambda^\epsilon \|\mathbf{u}_*\|_{2-1/q,q(\partial\Omega)} \right) \tag{3.28}$$

with $\epsilon = 1/3 - 2(1 - 1/q) > 0$. From Lemma 3.2 we know that there is $\lambda_1 > 0$ such that

$$|\mathcal{T}(\mathbf{u})| \leq c_{22} |\log \lambda|^{-1} \|\mathbf{u}_*\|_{2-1/q,q(\partial\Omega)}$$

for all $\lambda \in (0, \lambda_1]$. Thus, replacing this estimate into (3.28), we recover (3.4) with $\lambda_0 = \min\{1, \lambda_1\}$ and the lemma is completely proved. ∎

The next result shows an estimate for solutions to the OSEEN problem with zero boundary data. The main feature of such a result is that the constant c which enters the estimate can be rendered independently of $\lambda \in (0, \lambda_0]$.

Lemma 3.4. Let Ω be as in Lemma 3.2. Then, given

$$\mathbf{F} \in L^q(\Omega), \quad 1 < q < 6/5,$$

there exists at least one solution to the Oseen problem:

$$\left. \begin{array}{l} \Delta\mathbf{w} - \lambda\dfrac{\partial\mathbf{w}}{\partial x_1} = \nabla\tau + \mathbf{F} \\[2mm] \nabla \cdot \mathbf{w} = 0 \\[2mm] \mathbf{w} = 0 \ \text{at} \ \partial\Omega \\[2mm] \lim_{|x|\to\infty} \mathbf{w}(x) = 0. \end{array} \right\} \ \text{in} \ \Omega \tag{3.29}$$

such that

$$\mathbf{w} \in \mathcal{C}_q, \quad \tau \in D^{1,q}(\Omega).$$

This solution satisfies the estimate

$$\langle \mathbf{w} \rangle_{\lambda,q} \leq c \|\mathbf{F}\|_q \tag{3.30}$$

for all $\lambda \in (0, \lambda_0]$ (λ_0 given in Lemma 3.3) and with $c = c(\Omega, q, \lambda_0)$. Moreover, if $\mathbf{w}_1$, τ_1 is another solution to (3.29) corresponding to the same data and with $\langle \mathbf{w}_1 \rangle_{\lambda,q}$ finite[11], then $\mathbf{w} = \mathbf{w}_1$.

PROOF :   Extend $\mathbf{F}$ to zero outside Ω and continue to denote by $\mathbf{F}$ this extension. We look for a solution $\mathbf{w}$, τ of the form

$$\mathbf{w} = \mathbf{v} + \mathbf{u}, \quad \tau = p + \pi \tag{3.31}$$

where

$$\left. \begin{aligned} \Delta \mathbf{v} - \lambda \frac{\partial \mathbf{v}}{\partial x_1} &= \nabla p + \mathbf{F} \\ \nabla \cdot \mathbf{v} &= 0 \end{aligned} \right\} \quad \text{in } \mathbb{R}^2$$

while $\mathbf{u}$, π solve problem (3.3) with $\mathbf{u}_* = -\mathbf{v}$ at $\partial\Omega$. From Lemma 3.1 we know that there is a unique solution $\mathbf{v}$, p satisfying the estimate

$$\langle \mathbf{v} \rangle_{\lambda,q} + |\mathbf{v}|_{2,q} + |p|_{1,q} \le c_1 \|\mathbf{F}\|_q \tag{3.32}$$

with c_1 independent of $\lambda \in (0, 1]$. We need another estimate for a suitable norm of $\mathbf{v}$ on the unit ball B centered at the origin. To this end, we write $\mathbf{v}$ as the OSEEN volume potential[12]

$$\mathbf{v}(x) = \int_{\mathbb{R}^2} \mathbf{E}(x - y; \lambda) \cdot \mathbf{F}(y) dy = \int_{\mathbb{R}^2} \mathbf{E}(\lambda(x - y); 1) \cdot \mathbf{F}(y) dy$$

where the homogeneity property (2.8) has been employed. By HÖLDER inequality we have (with $q' = q/(q-1)$)

$$|\mathbf{v}(x)|^{q'} \le \left(\int_{\mathbb{R}^2} |\mathbf{E}(\lambda(x - y); 1)|^{q'} dy \right) \|\mathbf{F}\|_q^{q'}.$$

Since

$$\int_{\mathbb{R}^2} |\mathbf{E}(\lambda(x - y); 1)|^{q'} dy = \lambda^{-2} \int_{\mathbb{R}^2} |\mathbf{E}(z; 1)|^{q'} dz = c\lambda^{-2},$$

it follows

$$\|\mathbf{v}\|_{q', B_1} \le 2\pi c \lambda^{-2(1-1/q)} \|\mathbf{F}\|_q, \tag{3.33}$$

which is the inequality we wanted to show. We now pass to the solution $\mathbf{u}$, π. From Lemma 3.3 we know that

$$\mathbf{u} \in \mathcal{C}_q, \quad \pi \in D^{1,q}(\Omega) \tag{3.34}$$

---

[11] Of course, $\mathbf{w}_1$, τ_1 satisfy (3.29) in the sense of distribution.

[12] This is easily shown by a standard approximating procedure which starts with $\mathbf{F} \in C_0^\infty(\Omega)$ and uses (3.32).

and that, in addition, for all $\lambda \in (0, \lambda_0]$

$$\langle \mathbf{u} \rangle_{\lambda,q} \leq c_2 \lambda^{2(1-1/q)} \| \mathbf{v} \|_{2-1/q,q(\partial\Omega)}.$$

Our task is to increase the right hand side of (3.34) in terms of $\mathbf{F}$. To this end, we observe that, by the trace theorem of GAGLIARDO, by EHRLING and HÖLDER inequalities, it is[13]

$$\| \mathbf{v} \|_{2-1/q,q(\partial\Omega)} \leq c_3 \| \mathbf{v} \|_{2,q,\Omega_1} \leq c_4 \left(\| \mathbf{v} \|_{q,\Omega_1} + |\mathbf{v}|_{2,q,\Omega_1} \right)$$

with c_5 depending only on Ω. Being $q' > q$, from this latter inequality and from $(3.32) - (3.34)$ we conclude, for all $\lambda \in (0, \lambda_0]$

$$\langle \mathbf{u} \rangle_{\lambda,q} \leq c_5 \| \mathbf{F} \|_q \qquad (3.35)$$

with $c_5 = c_5(\Omega, q, \lambda_0)$. Estimate (3.30) becomes then a consequence of (3.31), (3.32) and (3.35). To show the result completely we have to prove the uniqueness part. However, this follows from Remark 2.1 and classical results for the OSEEN problem, *cf.*, CHANG & FINN, (1961, Theorems 4 and 6). ∎

Combining Lemmas 3.3 and 3.4, we immediately obtain the following result.

Lemma 3.5. *Let Ω be as in Lemma 3.2. Then, given*

$$\mathbf{F} \in L^q(\Omega), \quad \mathbf{u}_* \in W^{2-1/q,q}(\partial\Omega), \ 1 < q < 6/5,$$

there exists a unique solution to the Oseen problem

$$\left. \begin{array}{c} \Delta \mathbf{u} - \lambda \dfrac{\partial \mathbf{u}}{\partial x_1} = \nabla \pi + \mathbf{F} \\[2mm] \nabla \cdot \mathbf{u} = 0 \end{array} \right\} \ \textit{in} \ \Omega$$

$$\mathbf{u} = \mathbf{u}_* \ \ \textit{at} \ \ \partial\Omega$$

$$\lim_{|x| \to \infty} \mathbf{u}(x) = 0.$$

such that $\mathbf{u} \in \mathcal{C}_q$, $\pi \in D^{1,q}$. Moreover, there is $\lambda_0 > 0$ such that for all $\lambda \in (0, \lambda_0]$, this solution satisfies

$$\langle \mathbf{u} \rangle_{\lambda,q} \leq c \left[\lambda^{2(1-1/q)} |\log \lambda|^{-1} \| \mathbf{u}_* \|_{2-1/q,q(\partial\Omega)} + \| \mathbf{F} \|_q \right],$$

with $c = c(\Omega, q, \lambda_0)$.

Finally, we need the following result concerning functions having finite $\langle \cdot \rangle_{\lambda,q}-$ norm.

---

[13] We may assume, without loss, that Ω^c is contained in $B_{1/2}$.

19

Lemma 3.6. *Let Ω be an arbitrary domain in $\mathbb{R}^2$ and let $\mathbf{v}$, $\mathbf{w}$ be two solenoidal vector functions in Ω for which the norm (2.1), with $1 < q \le 6/5$, is finite. Then, the following inequality holds for all $\lambda > 0$*

$$\|\mathbf{v} \cdot \nabla \mathbf{w}\|_q \le 4\lambda^{-1-2(1-1/q)} \langle \mathbf{v} \rangle_{\lambda,q} \langle \mathbf{w} \rangle_{\lambda,q}.$$

PROOF : Taking into account that $\mathbf{v}$ and $\mathbf{w}$ are divergence free, we obtain

$$\mathbf{v} \cdot \nabla \mathbf{w} = \left(-v_1 \frac{\partial w_2}{\partial x_2} + v_2 \frac{\partial w_1}{\partial x_2} \right) \mathbf{e}_1 + \left(-v_1 \frac{\partial w_2}{\partial x_1} + v_2 \frac{\partial w_2}{\partial x_2} \right) \mathbf{e}_2$$

and so, by HÖLDER inequality and by (2.1),

$$
\begin{aligned}
\|\mathbf{v} \cdot \nabla \mathbf{w}\|_q &\le \left(\|v_1\|_{3q/(3-2q)} |w_2|_{1,3/2} + \|v_2\|_3 |\mathbf{w}|_{1,3q/(3-q)} \right) \\
&\le \left(\lambda^{-2/3} |w_2|_{1,3/2} \langle \mathbf{v} \rangle_{\lambda,q} + \lambda^{-1/3} \|v_2\|_3 \langle \mathbf{w} \rangle_{\lambda,q} \right)
\end{aligned}
\tag{3.36}
$$

From L^q-interpolation inequality we find (with $q' = q/(q-1)$)

$$|w_2|_{1,3/2} \le |w_2|_{1,q}^{3/q'} |w_2|_{1,3q/(3-q)}^{1-3/q'} \le \lambda^{-2/q'-1/3} \langle \mathbf{w} \rangle_{\lambda,q}$$

$$\|v_2\|_3 \le \|v_2\|_{2q/(2-q)}^{6/q'} \|\mathbf{v}\|_{1,3q/(3-2q)}^{1-6/q'} \le \lambda^{-2/q'-2/3} \langle \mathbf{v} \rangle_{\lambda,q}$$

and the lemma becomes a consequence of these relations and of (3.36). ∎

4. Existence and Uniquenes for the Nonlinear Problem at Low Reynolds Number.

We are now in a position to prove an existence and uniqueness result for the nonlinear problem. We shall prove it when Ω is an exterior domain, leaving to the reader the (much simpler) case $\Omega = \mathbb{R}^2$.

Theorem 4.1. *Let $\Omega \subset \mathbb{R}^2$ be an exterior domain of class C^2 and let*

$$\mathbf{f} \in L^q(\Omega), \quad \mathbf{v}_* \in W^{2-1/q,q}(\partial\Omega), \quad \mathbf{v}_\infty = \mathbf{e}_1,$$

with $q \in (1, 6/5)$. There exists a positive constant $\lambda_0 > 0$ such that if for some $\lambda \in (0, \lambda_0]$ it is

$$|\log \lambda|^{-1} \|\mathbf{v}_* - \mathbf{e}_1\|_{2-1/q,q(\partial\Omega)} + \lambda^{2/q-2} \|\mathbf{f}\|_q < 1/32c^2, \tag{4.1}$$

with c given in Lemma 3.5, then problem (1.1), (1.2), (1.4) has at least one solution $\mathbf{v}$, p such that

$$v_2 \in L^{2q/(2-q)}(\Omega) \cap D^{1,q}(\Omega)$$

$$\mathbf{v} \in L^{q/(3-2q)}(\Omega) \cap D^{1,3q/(3-q)}(\Omega) \cap D^{2,q}(\Omega)$$

$$p \in D^{1,q}(\Omega).$$

Furthermore, this solution satisfies the estimate

$$\langle \mathbf{v} - \mathbf{e}_1 \rangle_{\lambda,q} \le \frac{3}{2} c \left(\lambda^{2(1-1/q)} |\log \lambda|^{-1} \|\mathbf{v}_* - \mathbf{e}_1\|_{2-1/q,q(\partial\Omega)} + \|\mathbf{f}\|_q \right) \tag{4.2}$$

where $\langle \cdot \rangle_{\lambda,q}$ *is defined in (2.1). Finally, if* $\mathbf{v}_1, p_1$ *is another solution to problem (1.1), (1.2), (1.4) corresponding to the same data and such that*

$$c\lambda^{-2(1-1/q)} \langle \mathbf{v}_1 - \mathbf{e}_1 \rangle_{\lambda,q} < 13/64, \tag{4.3}$$

then $\mathbf{v} = \mathbf{v}_1$ *and* $p = p_1 + const.$

PROOF : We look for a solution $\mathbf{v}$ to (1.1), (1.2), (1.4) of the form

$$\mathbf{v} = \mathbf{u}_0 + \lambda^{2(1-1/q)} \mathbf{u} + \mathbf{e}_1, \qquad p = p_0 + \lambda^{2(1-1/q)} \pi, \tag{4.4}$$

where

$$\left. \begin{aligned} \Delta \mathbf{u}_0 - \lambda \frac{\partial \mathbf{u}_0}{\partial x_1} &= \nabla p_0 + \mathbf{f} \\ \nabla \cdot \mathbf{u}_0 &= 0 \end{aligned} \right\} \quad \text{in } \Omega \\ \mathbf{u}_0 = \mathbf{v}_* - \mathbf{e}_1 \text{ at } \partial\Omega \\ \lim_{|x|\to\infty} \mathbf{u}_0(x) = 0, \tag{4.5}$$

and

$$\left. \begin{aligned} \Delta \mathbf{u} - \lambda \frac{\partial \mathbf{u}}{\partial x_1} &= \lambda[\lambda^{2(1-1/q)} \mathbf{u} \cdot \nabla \mathbf{u} + \lambda^{2(1-1/q)} \mathbf{u}_0 \cdot \nabla \mathbf{u}_0 \\ &\quad + \mathbf{u}_0 \cdot \nabla \mathbf{u} + \mathbf{u} \cdot \nabla \mathbf{u}_0] + \nabla \pi \\ \nabla \cdot \mathbf{u} &= 0 \end{aligned} \right\} \quad \text{in } \Omega \\ \mathbf{u} = 0 \quad \text{at } \partial\Omega \\ \lim_{|x|\to\infty} \mathbf{u}(x) = 0. \tag{4.6}$$

A solution to (4.5) is determined by Lemma 3.5. For all $\lambda \in (0, \lambda_0]$ such a solution obeys the estimate

$$\langle \mathbf{u}_0 \rangle_{\lambda,q} \le c \left(\lambda^{2(1-1/q)} |\log \lambda|^{-1} \|\mathbf{u}_*\|_{2-1/q,q(\partial\Omega)} + \|\mathbf{f}\|_q \right) \equiv D. \tag{4.7}$$

A solution to (4.6) is likewise obtained from Lemma 3.3 with the help of the contraction mapping theorem. To this end, for $\lambda \in (0, \lambda_0]$ set

$$\mathcal{B}_{\lambda,q} = \left\{ \mathbf{w} : \Omega \to \mathbb{R}^2 : \langle \mathbf{w} \rangle_{\lambda,q} < \infty; \ \nabla \cdot \mathbf{w} = 0 \text{ in } \Omega; \ \mathbf{w} = 0 \text{ at } \partial\Omega \right\}.$$

Clearly, $\mathcal{B}_{\lambda,q}$ is a Banach space endowed with the norm $\langle\cdot\rangle_{\lambda,q}$. For $\delta > 0$, let

$$X_{\lambda,q,\delta} = \{\mathbf{w} \in \mathcal{B}_{\lambda,q} : \langle\mathbf{w}\rangle_{\lambda,q} \leq \delta\}.$$

Consider the mapping

$$L : \mathbf{w} \in X_{\lambda,q,\delta} \to L(\mathbf{w}) = \mathbf{u} \in \mathcal{B}_{\lambda,q}, \qquad (4.8)$$

where $\mathbf{u}$ solves

$$(\nabla\mathbf{u}, \nabla\boldsymbol{\varphi}) = -\lambda(\frac{\partial\mathbf{u}}{\partial x_1}, \nabla\boldsymbol{\varphi}) - (\mathcal{F}(\mathbf{w}), \boldsymbol{\varphi}) \qquad (4.9)_1$$

for all $\boldsymbol{\varphi} \in \mathcal{D}(\Omega)$, with

$$\mathcal{F}(\mathbf{w}) = \lambda\left(\lambda^{2(1-1/q)}\mathbf{w}\cdot\nabla\mathbf{w} + \lambda^{-2(1-1/q)}\mathbf{u}_0\cdot\nabla\mathbf{u}_0 + \mathbf{u}_0\cdot\nabla\mathbf{w} + \mathbf{w}\cdot\nabla\mathbf{u}_0\right). \quad (4.9)_2$$

By Lemma 3.6 it follows

$$\|\mathcal{F}(\mathbf{w})\|_q < \infty,$$

and so, by Lemma 3.4, we derive that the mapping (4.8) is well-defined and that, in addition, the solution $\mathbf{u}$ verifies

$$\mathbf{u} \in D^{2,q}(\Omega), \quad \pi \in D^{1,q}(\Omega) \qquad (4.10)$$

where π is the pressure field associated to $\mathbf{u}$. Thus, in particular, $\mathbf{u}(x)$ tends to zero uniformly as $|x| \to \infty$, cf. Remark 2.1. We next show that, given $\lambda \in (0, \lambda_0]$ there exists $\delta = \delta(\lambda) > 0$ such that L is a contraction in $X_{\lambda,q,\delta}$, provided the data obey condition (4.1). From Lemma 3.6 and from (4.7), $(4.9)_2$ we find

$$\|\mathcal{F}(\mathbf{w})\|_q \leq 4\left(\langle\mathbf{w}\rangle_{\lambda,q}^2 + \lambda^{-4(1-1/q)}D^2 + 2\lambda^{-2(1-1/q)}D\langle\mathbf{w}\rangle_{\lambda,q}\right)$$

and therefore, in view of Lemma 3.5, we obtain that for any $\mathbf{w} \in X_{\lambda,q,\delta}$ there is a unique solution $\mathbf{u}$ to (4.9) satisfying (4.10) and such that

$$\langle\mathbf{u}\rangle_{\lambda,q} = \langle L(\mathbf{w})\rangle_{\lambda,q} \leq 4c\left(\langle\mathbf{w}\rangle_{\lambda,q}^2 + \lambda^{-4(1-1/q)}D^2 + 2\lambda^{-2(1-1/q)}D\langle\mathbf{w}\rangle_{\lambda,q}\right). \quad (4.11)$$

Since $\langle\mathbf{w}\rangle_{\lambda,q} \leq \delta$, the preceding inequality yields

$$\langle L(\mathbf{w})\rangle_{\lambda,q} \leq 4c\left(\delta^2 + \lambda^{-4(1-1/q)}D^2 + 2\lambda^{-2(1-1/q)}D\delta\right). \qquad (4.12)$$

Thus, taking (for instance)

$$\delta \equiv \lambda^{-2(1-1/q)}D < (32c)^{-1},$$

from (4.11) it follows for $\lambda \in (0, \lambda_0]$

$$\langle L(\mathbf{w}) \rangle_{\lambda, q} \leq 4c \cdot \frac{\delta}{8c} = \frac{\delta}{2}$$

which furnishes that the mapping (4.8) transforms $X_{\lambda, q, \delta}$ into itself, with δ given in (4.13). It is now simple to show that, in fact, L is a contraction in $X_{\lambda, q, \delta}$. Actually, performing the same kind of reasoning leading to (4.11), for any $\mathbf{w}_1, \mathbf{w}_2 \in X_{\lambda, q, \delta}$ we show

$$\langle L(\mathbf{w}_1) - L(\mathbf{w}_2) \rangle_{\lambda, q} \leq 4c[2\lambda^{-2(1-1/q)} \langle \mathbf{u}_0 \rangle_{\lambda, q} \langle \mathbf{w}_1 - \mathbf{w}_2 \rangle_{\lambda, q}$$
$$+ (\langle \mathbf{w}_1 \rangle_{\lambda, q} + \langle \mathbf{w}_2 \rangle_{\lambda, q}) \langle \mathbf{w}_1 - \mathbf{w}_2 \rangle_{\lambda, q}]$$

and so, by (4.7) and (4.13) we deduce

$$\langle L(\mathbf{w}_1 - L(\mathbf{w}_2) \rangle_{\lambda, q} \leq 8c \left(\lambda^{-2(1-1/q)} D + \delta \right) \langle \mathbf{w}_1 - \mathbf{w}_2 \rangle_{\lambda, q} \leq \frac{1}{2} \langle \mathbf{w}_1 - \mathbf{w}_2 \rangle_{\lambda, q}$$

which proves that L is a contraction in $X_{\lambda, q, \delta}$. We may thus conclude that, under the assumption (4.13) (that is, (4.1)) on $\mathbf{v}_*$, $\mathbf{f}$ and λ, problem (4.6) admits at least one solution $\mathbf{u}$ with $\langle \mathbf{u} \rangle_{\lambda, q}$ finite. In fact, in view of Lemma 3.4, it follows that $\mathbf{u}$, π satisfy also condition (4.10) with π pressure field associated to $\mathbf{u}$. As a consequence, the fields (4.4) constitute a solution to (1.1), (1.2), (1.4). Furthermore, by (4.7), (4.13) and (4.14) we infer that $\mathbf{v}$ satisfies also estimate (4.2). It remains to show uniqueness. Setting

$$\mathbf{u} = \mathbf{v} - \mathbf{e}_1, \ \mathbf{u}_1 = \mathbf{v}_1 - \mathbf{e}_1$$
$$\mathbf{w} = \mathbf{u} - \mathbf{u}_1, \ \pi = p - p_1,$$

we have

$$\left. \begin{aligned} \Delta \mathbf{w} - \lambda \frac{\partial \mathbf{w}}{\partial x_1} &= \nabla \pi + \mathbf{F} \\ \nabla \cdot \mathbf{w} &= 0 \end{aligned} \right\} \quad \text{in } \Omega$$
$$\mathbf{w} = 0 \quad \text{at} \ \partial \Omega$$
$$\lim_{|x| \to \infty} \mathbf{w}(x) = 0,$$

where

$$\mathbf{F} = \lambda \left(\mathbf{w} \cdot \nabla \mathbf{u}_1 + \mathbf{u} \cdot \nabla \mathbf{w} \right).$$

From Lemmas 3.5 and 3.6 it follows that

$$\langle \mathbf{w} \rangle_{\lambda, q} \leq 4c\lambda^{-2(1-1/q)} \langle \mathbf{w} \rangle_{\lambda, q} \left(\langle \mathbf{u}_1 \rangle_{\lambda, q} + \langle \mathbf{u} \rangle_{\lambda, q} \right).$$

By a direct computation which makes use of $(4.1) - (4.3)$ we find

$$4c\lambda^{-2(1-1/q)} \left(\langle \mathbf{u}_1 \rangle_{\lambda, q} + \langle \mathbf{u} \rangle_{\lambda, q} \right) < 1$$

so that (4.15) implies $\mathbf{w} \equiv 0$, thus completing the proof of the theorem. ∎

Remark 4.1. Theorem 4.1 does not require that the datum $\mathbf{v}_*$ satisfies the zero-outflux condition:

$$\int_{\partial\Omega} \mathbf{v}_* \cdot \mathbf{n} = 0.$$

Remark 4.2. Solutions determined in Theorem 4.1 are unique in the class of those solutions verifying (4.3). This result is much weaker than the analogous one proved for the three-dimensional case. In fact, for the situation at hand, we must require that *both* $\mathbf{v}$ *and* $\mathbf{v}_1$ are small in suitable norms and it is not known if a solution $\mathbf{v}$ obeying (4.2) is unique in the class of those solutions $\mathbf{v}_1$ which *merely* satisfy the condition $\langle \mathbf{v}_1 \rangle_{\lambda,q} < \infty$.

Remark 4.3. Since $q < 6/5$, solutions determined in Theorem 4.1 belong to $D^{1,s}(\Omega)$, for some $s < 2$. Moreover, they are also D-solutions in the sense that they have a finite Dirichlet integral. In fact, since $\mathbf{v} \in D^{2,q}(\Omega) \cap D^{1,3q/(3-q)}(\Omega)$, $1 < q \leq 6/5$, by Lemma 2.1 it follows $\mathbf{v} \in D^{1,2q/(2-q)}(\Omega)$ and so, noticing that

$$3q/(3-q) \leq 2 < 2q/(2-q)$$

we conclude, by interpolation, $\mathbf{v} \in D^{1,2}(\Omega)$.

5. Limit of Vanishing Reynolds Number: Transition to the Stokes Problem

In this section we shall investigate the behaviour of solutions constructed in Theorem 4.1, in the limit of vanishing Reynolds number. We shall prove, in particular, that they converge to a uniquely determined solution of a suitable Stokes problem. However, this limiting process does not need to preserve the condition at infinity on the velocity field and, in fact, this condition is preserved *if and only if* the data are prescribed in a certain way.

For the sake of simplicity, we assume throughout that the body force $\mathbf{f}$ is identically vanishing. Generalizations are left to the reader.

The first step is to show a uniform bound (independent of λ) for solutions determined in Theorem 4.1. This will be established in the following

Lemma 5.1. *Let the assumptions of Theorem 4.1 be satisfied and let $\mathbf{f} \equiv 0$. Then, there exists $c = c(\Omega, q, \mathbf{v}_*, \lambda_0) > 0$ such that*

$$\int_\Omega \nabla\mathbf{v} : \nabla\mathbf{v} \leq c.$$

PROOF : We suppose, as usual, the origin of coordinates in the interior of Ω^c. Set

$$\sigma = m\nabla \log |x|$$

24

with m such that

$$\int_{\partial\Omega} (\mathbf{v}_* - \boldsymbol{\sigma}) \cdot \mathbf{n} = 0.$$

In view of this latter property, by known results on the extension of boundary data to solenoidal fields, $cf.$, $e.g.$, GALDI (1993, Chapter II, Section 3), there exists a vector field $\mathbf{V}$ satisfying

    (i) $\nabla \cdot \mathbf{V} = 0$ in Ω,
    (ii) $\mathbf{V} = \mathbf{v}_* - \boldsymbol{\sigma}$ at ,$\partial\Omega$,
    (iii) $\mathbf{V} \in W^{1,2}(\Omega)$,
    (iv) $\mathbf{V} \equiv 0$ in Ω^R, for some $R > \delta(\Omega^c)$.

Setting

$$\mathbf{v} = \mathbf{w} + \mathbf{e}_1 + \boldsymbol{\sigma} + \mathbf{V}, \tag{5.1}$$

from the property of $\mathbf{v}$, $\boldsymbol{\sigma}$ and $\mathbf{V}$ and with the help of the embedding theorems, we deduce that the field $\mathbf{w}$ verifies the following properties

$$\langle \mathbf{w} \rangle_{\lambda,q} < \infty$$
$$\mathbf{w} = 0 \text{ at } \partial\Omega \tag{5.2}$$
$$\lim_{|x| \to \infty} \mathbf{w}(x) = 0.$$

We multiply $(1.1)_1$ by $\mathbf{v}$ and integrate over Ω. By virtue of (5.1), (5.2) and of the summability properties of $\mathbf{v}$, performing an easily justified integration by parts which uses the relation

$$\frac{\partial \sigma_i}{\partial x_j} = \frac{\partial \sigma_j}{\partial x_i} \quad i,j = 1,2$$

we find

$$|\mathbf{w}|_{1,2}^2 = \int_\Omega \left\{ \nabla\mathbf{w} : \nabla\mathbf{V} - \lambda[\mathbf{w} \cdot \nabla\mathbf{w} \cdot (\boldsymbol{\sigma} + \mathbf{V}) + \mathbf{V} \cdot \nabla\mathbf{w} \cdot \mathbf{V} + \mathbf{V} \cdot \frac{\partial \mathbf{w}}{\partial x_1}] \right\}. \tag{5.3}$$

By the properties of $\mathbf{V}$ and again by the embedding theorems, we have

$$\|\mathbf{V}\|_r < \infty \quad \text{for all } r \in [1, \infty).$$

Therefore, using in (5.3) the SCHWARZ and CAUCHY inequalities allows us to show

$$|\mathbf{w}|_{1,2}^2 \leq \lambda \left| \int_\Omega \mathbf{w} \cdot \nabla\mathbf{w} \cdot (\boldsymbol{\sigma} + \mathbf{V}) \right| + c_1 \tag{5.4}$$

where c_1 depends on $\mathbf{V}$, Ω and λ_0 but not on λ. We now observe that, by HÖLDER inequality, it follows

$$\left| \int_\Omega \mathbf{w} \cdot \nabla\mathbf{w} \cdot (\boldsymbol{\sigma} + \mathbf{V}) \right| \leq \|\mathbf{w}\|_{3q/(3-2q)} |\mathbf{w}|_{1,3q/(3-q)} \|\boldsymbol{\sigma} + \mathbf{V}\|_{q/(2q-2)}$$

and so, being $q/(2q - 2) > 2$, from (5.4) and (2.1) we recover

$$\|\mathbf{w}\|_{1,2}^2 \leq c_2 \langle \mathbf{w} \rangle_{\lambda,q}^2 + c_1 \tag{5.5}$$

with c_2 independent of $\lambda \in (0, \lambda_0]$. From the estimate (4.2) and from (5.1) we then obtain, at once,

$$\langle \mathbf{w} \rangle_{\lambda,q} \leq c_3$$

with c_3 independent of $\lambda \in (0, \lambda_0]$. The lemma follows from this latter inequality, from (5.5) and from (5.1). ∎

The next lemma shows other uniform bounds for solutions on every compact set in $\bar{\Omega}$.

Lemma 5.2. *Let the assumptions of Lemma 5.1 be satisfied. Then, there exists a constant* $c = c(\Omega, R, q, \lambda_0, \mathbf{v}_*)$ *such that for all* $R > \delta(\Omega)^c$

$$\|\mathbf{v}\|_{2,q,\Omega_R} + \|p\|_{1,q,\Omega_R} \leq c.$$

PROOF : Using well–known estimates for the Stokes problem in bounded domains, *cf.*, *e.g.*, GALDI (1993, Chapter IV), from (1.1), (1.2) we derive

$$\|\mathbf{v}\|_{2,q,\Omega_R} + \|p\|_{1,q,\Omega_R} \leq c_1(\lambda\|\mathbf{v} \cdot \nabla \mathbf{v}\|_{q,\Omega_{R_1}} + \|\mathbf{v}\|_{1,q,\Omega_{R_1}} + \|p\|_{q,\Omega_{R_1}} + \|\mathbf{v}_*\|_{2-1/q,q(\partial\Omega)}). \tag{5.6}$$

for all $R_1 > R > \delta(\Omega^c)$ and with $c = c(\Omega, q, R, R_1)$. From the embedding theorems we easily deduce for any $s \in [1, \infty)$ and any $R > \delta(\Omega^c)$

$$\|\mathbf{v}\|_{s,\Omega_R} \leq c_2(\|\mathbf{v}\|_{2,\Omega_R} + |\mathbf{v}|_{1,2,\Omega_R})$$
$$\leq c_3(|\mathbf{v}|_{1,2} + \|\mathbf{v}_*\|_{2-1/q,q(\partial\Omega)})$$

and so, by Lemma 5.1,

$$\|\mathbf{v}\|_{s,\Omega_R} \leq c_4, \quad s \geq 1, \ R > \delta(\Omega^c) \tag{5.7}$$

with $c_4 = c_4(\Omega, R, s)$. Since

$$\|\mathbf{v} \cdot \nabla \mathbf{v}\|_{q,\Omega_{R_1}} \leq \|\mathbf{v}\|_{2q/(2-q),\Omega_{R_1}} |\mathbf{v}|_{1,2,\Omega_{R_1}},$$

coupling (5.6) and (5.7), with the help of Lemma 5.1 it follows

$$\|\mathbf{v}\|_{2,q,\Omega_R} + \|p\|_{1,q,\Omega_R} \leq c_5(\|p\|_{q,\Omega_{R_1}} + 1) \tag{5.8}$$

with c_5 independent of $\lambda \in (0, \lambda_0]$. To estimate the pressure term at the right hand side of (5.8), we use the following inequality (obtained after possibly modifying p by addition of a constant)

$$\|p\|_{q,\Omega_{R_1}} \leq c_6(\lambda\|\mathbf{v}\|_{2q,\Omega_{R_1}}^2 + |\mathbf{v}|_{1,2})$$

which, by (5.7) and Lemma 5.1, in turn implies

$$\|p\|_{q,\Omega_{R_1}} \leq c_7,$$

with c_7 independent of $\lambda \in (0, \lambda_0]$. The lemma then follows from this latter inequality and from (5.8). ∎

The next result shows that solutions of Theorem 4.1 tend, in the limit $\lambda \to 0$, to solutions of a suitable Stokes problem.

Lemma 5.3. *Let the assumptions of Lemma 5.1 be satisfied. Denote by* $\mathbf{w}$, π *the unique solution to the following Stokes problem*

$$\left.\begin{array}{c} \Delta \mathbf{w} = \nabla \pi \\ \nabla \cdot \mathbf{w} = 0 \end{array}\right\} \ \ in \ \ \Omega$$

$$\mathbf{w} = \mathbf{v}_* \ \ at \ \ \partial\Omega \tag{5.9}$$

$$|\mathbf{w}|_{1,2} < \infty.$$

Then, as $\lambda \to 0$, *the solutions* $\mathbf{v}$, p *constructed in Theorem 4.1 satisfy*

$$\mathbf{v} \to \mathbf{w} \quad weakly \ in \ \ D^{1,2}(\Omega)$$

$$\mathbf{v} \to \mathbf{w} \quad weakly \ in \ \ W^{2,q}(\Omega_R) \tag{5.10}$$

$$p \to \pi \quad weakly \ in \ \ W^{1,q}(\Omega_R),$$

for any $R > \delta(\Omega^c)$.

PROOF : Let $\{\lambda_n\}_{n \in \mathbb{N}}$ be any sequence converging to zero. From Lemmas 5.1 and 5.2 it readily follows that, along a subsequence at least, conditions (5.10) hold. However, from known results on the Stokes problem, *cf.* CHANG & FINN (1961, Theorems 1 and 6), the field $\mathbf{w}$ satisfying (5.9) is uniquely determined, whatever the sequence may be, and, therefore, the result is proved.

Before stating the main result of this section, we should recall a result due to GALDI & SIMADER (1990, Theorem 4.1) which asserts that solutions $\mathbf{u}$, τ to the following homogeneous Stokes system

$$\left.\begin{array}{c} \Delta \mathbf{u} = \nabla \tau \\ \nabla \cdot \mathbf{u} = 0 \end{array}\right\} \ in \Omega$$

$$\mathbf{u} = 0 \ \ at \ \partial\Omega \tag{5.11}$$

$$\mathbf{u} \in D_0^{1,r}(\Omega), \ \ r > 2,$$

constitute a two-dimensional vector space $\mathcal{S}$. A basis in $\mathcal{S}$ is denoted by $\mathbf{h}^{(i)}$, $\pi^{(i)}$, $i = 1, 2$. These basis–solutions will be called "exceptional". In the case when Ω is the exterior of a unit circle we can give an explicit form to the exceptional solutions and we have

$$h_1^{(1)} = 2\log|x| + 2x_2^2/|x|^2 + (x_1^2 - x_2^2)/|x|^4 - 1,$$

$$h_2^{(1)} = 4x_1x_2\log|x|^2/|x|^2,$$

$$\pi^{(1)} = 4x_1/|x|^2,$$

$$h_1^{(2)} = 2\log|x| + 2x_1^2/|x|^2 + (x_2^2 - x_1^2)/|x|^4 - 1, \tag{5.12}$$

$$h_2^{(2)} = 4x_1x_2\log|x|^2/|x|^2,$$

$$\pi^{(2)} = 4x_2/|x|^2.$$

We are now in a position to show the following main result.

Theorem 5.1. *Let the assumptions of Theorem 4.1 hold and let* $\mathbf{v}$, p *be the solution constructed in that theorem, corresponding to* $\mathbf{f} \equiv 0$. *Then, denoting by* $\mathbf{w}$, π *the solution to the Stokes problem (5.9) we have that, as* $\lambda \to 0$, $\mathbf{v}$, p *tend to* $\mathbf{w}$, π *in the sense specified by Lemma 5.3. Moreover, there is* $\mathbf{w}_0 \in \mathbb{R}^2$ *such that*

$$\lim_{|x| \to \infty} \mathbf{w}(x) = \mathbf{w}_0 \qquad (5.13)_1$$

and we have

$$\mathbf{w}_0 - \mathbf{e}_1 = -\frac{1}{4\pi} \lim_{\lambda \to 0} \boldsymbol{T}(\mathbf{v})|\log \lambda| \qquad (5.13)_2$$

where

$$\boldsymbol{T}(\mathbf{v}) = -\int_{\partial\Omega} \mathbf{T}(\mathbf{v}, p) \cdot \mathbf{n}.$$

Finally, the limit process preserves the prescription at infinity, i.e., $\mathbf{w}_0 = \mathbf{e}_1$ *if and only if the data satisfy the conditions*

$$\int_{\partial\Omega} (\mathbf{v}_* - \mathbf{e}_1) \cdot \mathbf{T}(\mathbf{h}^{(i)}, \pi^{(i)}) \cdot \mathbf{n} = 0 \quad i = 1, 2, \qquad (5.14)$$

where $\{\mathbf{h}^{(i)}, \pi^{(i)}\}$ *are the "exceptional" solutions to the homogeneous Stokes system (5.11). In the particular case where* $\partial\Omega$ *is a unit circle, (5.14) reduces to*

$$\int_{\partial\Omega} (v_{*i} - \delta_{1i}) = 0 \quad i = 1, 2.$$

PROOF : The first part of the statement follows from Lemma 5.3. Moreover, from classical results of CHANG & FINN (1961, Theorem 1) we know that $\mathbf{w}$ satisfies the representation

$$w_j = w_{0j} + \int_{\partial\Omega} [v_{*i} T_{il}(\mathbf{u}_j, e_j)(x - y) - U_{ij}(x - y) T_{il}(\mathbf{w}, \pi)(y)] n_l(y) d\sigma_y, \quad (5.15)$$

for some $\mathbf{w}_0 \in \mathbb{R}^2$, and with

$$\mathbf{u}_j = (U_{1j}, U_{2j}).$$

By the regularity of $\mathbf{v}_*$, (5.15) proves $(5.13)_1$. On the other hand, we have that $\mathbf{v}$ satisfies, *cf.* SMITH (1965, Theorem 6)

$$u_j(x) = \lambda \int_\Omega E_{ij}(x - y) u_l(y) D_l u_i(y) dy$$
$$+ \int_{\partial\Omega} [u_i(y) T_{il}(\mathbf{w}_j, e_j)(x - y) \qquad (5.16)$$
$$- E_{ij}(x - y) T_{il}(\mathbf{u}, p)(y) + \lambda u_i(y) E_{ij}(x - y) \delta_{1l}] n_l d\sigma_y,$$

where $\mathbf{u} = \mathbf{v} - \mathbf{e}_1$. We wish now to take $\lambda \to 0$ into (5.16). From HÖLDER inequality, from (2.1), (2.8) and (2.12) it follows for $q \in (1, 6/5)$

$$\left| \int_\Omega E_{ij}(x-y)u_l D_l u_i(y) dy \right| \leq \|\mathbf{u}\|_{3q/(3-2q)} |\mathbf{u}|_{1,3q/(3-q)} \|E\|_{q/(2q-q)}$$

$$\leq \lambda^{-1} \lambda^{-4(1-1/q)} \langle \mathbf{u} \rangle_{\lambda,q}^2.$$

where $c_1 = c_1(q, \Omega, \lambda_0)$. However, the solutions of Theorem 4.1 verify the estimate (4.1) and so, from the preceding inequality, we obtain

$$\lambda \left| \int_\Omega E_{ij}(x-y)u_l(y)D_l u_i(y) dy \right| \leq c_2 |\log \lambda|^{-1}, \tag{5.17}$$

where c_2 is independent of $\lambda \in (0, \lambda_0]$. Now, integrating by parts over the region Ω^c, we find

$$\int_{\partial\Omega} [T_{il}(\mathbf{w}_j, e_j(x-y)) + \lambda E_{ij}(x-y)\delta_{1l}] n_l(y) d\sigma_y = 0. \tag{5.18}$$

Moreover, by Lemma 5.2 and by the compact embedding theorems of trace at the boundary, cf. NEČAS (1967, Chapitre 2, §§5 and 6), it follows

$$\|\mathbf{v} - \mathbf{w}\|_{1,q,\partial\Omega} + \|p - \pi\|_{q,\partial\Omega} \to 0 \quad \text{as } \lambda \to 0. \tag{5.19}$$

With the aid of (5.16), (5.17) and of the asymptotic formula (2.7) we thus obtain

$$\lim_{\lambda \to 0} \int_{\partial\Omega} [u_i(y)T_{il}(\mathbf{w}_j, e_j)(x-y) + \lambda u_i(y)E_{ij}(x-y)\delta_{1l}] n_l d\sigma_y$$
$$= \int_{\partial\Omega} v_{*i}(y)T_{il}(\mathbf{u}_j, e_j)(x-y) d\sigma_y. \tag{5.20}$$

Finally, again from (2.7) and (5.17) we find

$$-\lim_{\lambda \to 0} \int_{\partial\Omega} E_{ij}(x-y)T_{il}(\mathbf{u}, p)(y) n_l(y) d\sigma_y$$
$$= \frac{1}{4\pi} \lim_{\lambda \to 0} \left[\int_{\partial\Omega} T_{il}(\mathbf{u}, p) n_l \right] |\log \lambda| \tag{5.21}$$
$$- \int_{\partial\Omega} U_{ij}(x-y)T_{il}(\mathbf{w}, \pi)(y) n_l(y) d\sigma_y.$$

Collecting (5.15)–(5.17), (5.20), (5.21) we obtain $(5.13)_2$. Let us show the last part of the theorem. We have $\mathbf{w}_0 = \mathbf{e}_1$ if and only if $\mathbf{u}_0 \equiv \mathbf{w} - \mathbf{e}_1$ is a solution to the problem

$$\left. \begin{array}{r} \Delta \mathbf{u}_0 = \nabla \pi \\[4pt] \nabla \cdot \mathbf{u}_0 = 0 \end{array} \right\} \quad \text{in } \Omega$$
$$\mathbf{u}_0 = \mathbf{v}_* - \mathbf{e}_1 \equiv \mathbf{u}_* \quad \text{at } \partial\Omega \tag{5.22}$$
$$\lim_{|x| \to \infty} \mathbf{u}_0(x) = 0.$$

Let us put this system into an equivalent form. We write

$$\mathbf{u}_0 = \mathbf{z} + \mathbf{V} + \boldsymbol{\sigma}$$

where $\boldsymbol{\sigma}$ is defined in the proof of Lemma 5.1 – with $\mathbf{u}_*$ in place of $\mathbf{v}_*$ – and $\mathbf{V}$ is a smooth solenoidal extension in Ω, of compact support, of the field $\mathbf{u}_* - \boldsymbol{\sigma}$[14]. Thus, (5.22) can be equivalently rewritten as follows

$$
\left.
\begin{aligned}
\Delta \mathbf{z} &= \nabla \pi + \mathbf{f} \\
\nabla \cdot \mathbf{z} &= 0
\end{aligned}
\right\} \text{ in } \Omega
$$

$$\mathbf{z} = 0 \text{ at } \partial\Omega \qquad (5.23)$$

$$\lim_{|x|\to\infty} \mathbf{z}(x) = 0,$$

with

$$\mathbf{f} = -\Delta\mathbf{V}.$$

Clearly, $\mathbf{f}$ defines a linear, bounded functional on $D_0^{1,q'}(\Omega)$, for all $q' > 1$, via the map

$$\varphi \in D_0^{1,q'}(\Omega) \to (\mathbf{f}, \varphi) \in \mathbb{R}.$$

Furthermore, in view of the properties of the exceptional solutions $\mathbf{h}^{(i)}, \pi^{(i)}, i = 1, 2$, and of $\boldsymbol{\sigma}$ we find

$$
\begin{aligned}
(\mathbf{f}, \mathbf{h}^{(i)}) &= -\int_\Omega \Delta\mathbf{V}\cdot\mathbf{h}^{(i)} = -\int_\Omega \Delta(\mathbf{V}+\boldsymbol{\sigma})\cdot\mathbf{h}^{(i)} \\
&= \int_{\partial\Omega} \mathbf{u}_* \cdot \mathbf{T}(\mathbf{h}^{(i)}, \pi^{(i)})\cdot\mathbf{n}, \ i = 1, 2.
\end{aligned}
\qquad (5.24)
$$

Assume now

$$\int_{\partial\Omega} \mathbf{u}_* \cdot \mathbf{T}(\mathbf{h}^{(i)}, \pi^{(i)})\cdot\mathbf{n} \equiv \int_{\partial\Omega} (\mathbf{v}_* - \mathbf{e}_1)\cdot\mathbf{T}(\mathbf{h}^{(i)}, \pi^{(i)})\cdot\mathbf{n} = 0. \ i = 1, 2. \qquad (5.25)$$

Then, by (5.24), $(\mathbf{f}, \mathbf{h}^{(i)}) = 0$, $i = 1, 2$ and by the results of GALDI & SIMADER (1990, Theorem 6.1) it follows that there exists a (unique) solution to (5.23) or, equivalently, to (5.22). Conversely, assume that there is a solution $\mathbf{u}_0$ to (5.22). Consequently, $\mathbf{z} = \mathbf{u}_0 - \mathbf{V} - \boldsymbol{\sigma}$ solves (5.23). From the work of CHANG & FINN (1961, Theorem 1) we know that $\mathbf{z}$ admits the following asymptotic estimate

$$\nabla\mathbf{z}(x) = O(|x|^{-2}), \quad \text{as } |x| \to \infty.$$

---

[14] The existence of $\mathbf{V}$ is guaranteed by well–known extension theorems, cf. GALDI (1993, Chapter III, Section 3).

From this relation and from the regularity near the boundary for the Stokes problem, *cf.* GALDI (1993, Chapter III, Theorem 5.1) it follows

$$\mathbf{z} \in D^{1,q}(\Omega), \quad \text{for all } q \in (1,2).$$

In addition, $\mathbf{z}$ vanishes at $\partial\Omega$ and at infinity, so that, by Lemma 2.2,

$$\mathbf{z} \in D_0^{1,q}(\Omega), \quad \text{for } q \in (1,2).$$

This condition together with the results of GALDI & SIMADER (1990, Theorem 6.1) implies that

$$(\mathbf{f}, \mathbf{h}^{(i)}) = 0, \quad i = 1, 2,$$

which, in turn, by (5.24), yields (5.25), thus proving the desired result. Finally, for Ω the exterior of a unit disc, replacing (5.12) into (5.25) one easily shows that (5.25) becomes

$$\int_{\partial\Omega} (\mathbf{v}_* - \mathbf{e}_1) = 0,$$

and the proof of the theorem is completed. ∎

Remark 5.1. An interesting consequence of Theorem 5.1 is the derivation of an asymptotic formula (in the limit of vanishing Reynolds number) for the force of "drag" $\mathcal{T}(\mathbf{v})$ exerted by the fluid on a body moving in it with constant velocity $\mathbf{e}_1$. Specifically, taking $\mathbf{v}_* \equiv 0$, from Theorem 5.1 we have that the limit solution $\mathbf{w}$ is identically zero so that from $(5.11)_2$ it easily follows in the limit $\lambda \to 0$

$$\mathcal{T}(\mathbf{v}) = (4\pi\mathbf{e}_1 + o(1)) |\log\lambda|.$$

This formula shows that in the limit of vanishingly small Reynolds number, the drag is determined *entirely* by the velocity at infinity $\mathbf{e}_1$ and that it is directed along the line of this vector. Surprisingly enough, it does *not* depend on the shape of the body. Equation (5.26) was obtained for the first time by FINN & SMITH (1967, Theorem 5.4). However, the class of solutions $\mathbf{v}$, p used by these authors differs *a priori* from those of Theorem 4.1.

Acknowledgement

This work was made under the auspices of G. N. F. M. of the Italian C. N. R. and with the financial support of M. P. I. 40% and 60% contracts at the University of Ferrara.

References

AMICK, C. J., 1988, On Leray's Problem of Steady Navier-Stokes Flow Past a Body in the Plane, *Acta Math.*, **161**, 71-130.

AMICK, C. J., 1991, On the Asymptotic Form of Navier-Stokes Flow Past a Body in the Plane, *J. Diff. Equations,* **91**, 149-167.

CHANG, I. D. & FINN, R., 1961, On the Solutions of a Class of Equations Occurring in Continuum Mechanics with Application to the Stokes Paradox, *Arch. Rational Mech. Anal.,* **7**, 388-341.

FINN, R. 1959, On Steady-State Solutions of the Navier-Stokes Partial Differential Equations, *Arch. Rational Mech. Anal.,* **3**, 381-396.

FINN, R. & SMITH, D.R., 1967, On the Stationary Solution of the Navier-Stokes Equations in Two Dimensions, *Arch. Rational Mech. Anal.,* **25**, 26-39.

GALDI, G.P., 1991, On the Existence of steady motions of a viscous flow with non-homogeneous boundary conditions, *Le Matematiche*, **66** (1), 503-524

GALDI, G.P., 1992, On the Oseen Boundary-Value Problem in Exterior Domains, *Springer Lecture Notes in Mathematics*, **1530**, 111-131.

GALDI, G.P., 1993, *An Introduction to the Mathematical Theory of the Navier-Stokes Equations, Vol. I: Linearized Stationary Problems*, Springer Tracts in Natural Phylosophy, in the Press.

GALDI, G.P. & SIMADER C.G., 1990, Existence, Uniqueness and L^q-estimates for the Stokes Problem in an Exterior Domain, *Arch. Rational Mech. Anal.,* **112**, 291-318.

GILBARG, D. & WEINBERGER, H.F., 1974, Asymptotic Properties of Leray's Solution of the Stationary Two-Dimensional Navier-Stokes Equations, *Russian Math. Surveys,* **29**, 109-123.

GILBARG, D. & WEINBERGER, H.F., 1978, Asymptotic Properties of Steady Plane Solutions of the Navier-Stokes Equations with Bounded Dirichlet Integral, *Ann. Scuola Norm. Sup. Pisa,* (4), **5**, 381-404.

KOZONO, H. & SOHR, H., 1991, New a priori Estimates for the Stokes Equations in Exterior Domains, *Indiana University Math. J.,* **40**, 1-28.

KOZONO, H. & SOHR, H., 1993, On a New Class of Generalized Solutions for the Stokes Equations in Exterior Domains, *Ann. Scuola Norm. Sup. Pisa,* (4), in the Press.

LERAY, J., 1933, Étude de Diverses Équations Intégrales non Linéaires et de Quelques Problèmes que Pose l'Hydrodynamique, *J. Math. Pures Appl.,* **12**, 1-82.

NEČAS, J., 1967, *Les Méthodes Directes en Théorie des Équations Elliptiques*, Masson et Cie.

OSEEN, C.W., 1927, *Neuere Methoden und Ergebnisse in der Hydrodynamik*, Leipzig, Akad. Verlagsgesellschaft M.B.H.

SMITH, D.R., 1965, Estimates at Infinity for Stationary Solutions of the Navier-Stokes Equations in Two Dimensions, *Arch. Rational Mech. Anal.*, **20**, 341-372.

Giovanni P. Galdi
Istituto di Ingegneria
Università di Ferrara
44100 Ferrara
Italy

VIVETTE GIRAULT

The Divergence, Curl and Stokes Operators in Exterior Domains of $\mathbb{R}^3$

1. An introduction to weighted Sobolev spaces

These series of lectures are adapted from two previous works of the author [12] and [13]. Their purpose is to show that weighted Sobolev spaces are a very convenient tool for studying the Stokes problems and its related divergence and curl operators in exterior domains of $\mathbb{R}^3$. What is meant by an exterior domain is the complement Ω of a bounded region $\overline{\Omega'}$ with a reasonably smooth boundary Γ. Ω' can be viewed as an obstacle and we are interested in flows past this obstacle. Of course, such problems are difficult because exterior domains are unbounded in all directions. To solve these problems, one cannot rely on classical results (as the Poincaré Inequality, for instance), but one has to take into account the behaviour of the data and solutions at infinity. It turns out that the weighted Sobolev spaces we shall use here describe accurately the growth or decay of functions at infinity. They were introduced and studied by Hanouzet in [21] and a wide range of basic elliptic problems were solved in these spaces by Giroire in [17].

Several results stated or established here were partially proved by other authors elsewhere. The most relevant references are those of Borchers & Sohr [4], Galdi & Simader [9], Giga & Sohr [11], Girault & Sequeira [15], Gunzburger [20], Heywood [23], Kozono & Sohr [24], Ladyzhenskaya & Solonnikov [25], Miyakawa [27] and von Wahl [34]. But the originality of these lectures is that the theorems that we prove are valid for a complete set of finite integer weights at infinity. Another interesting feature of this text is that its approach is fairly simple, although the results established look difficult at first sight. This is achieved by proving first the same results in the whole space $\mathbb{R}^3$ (and for this, the material of [17] is very handy) and afterwards by using the fact that these results have all been proved in a bounded domain.

We shall now describe briefly Hanouzet's weighted Sobolev spaces. The properties listed below were proved by Hanouzet [21] or Giroire [17]; the reader can refer to these two references for more details.

First, we specify some geometric characteristics of the domain. As far as the regularity of the boundary Γ is concerned, we shall always assume that it is at least Lipschitz-continuous and we shall specify whenever we require more regularity. The reader can refer to the definitions in [18] for all that concerns the regularity of the boundary. Unless otherwise specified, Ω' will not be assumed to be connected; we suppose that it has I connected components that we denote by Ω_i', for $1 \leq i \leq I$, and that each of them has a connected boundary Γ_i. This guarantees that its complement

Ω is connected. Let $\mathbf{x} = (x_1, x_2, x_3)$ denote a general point of $\mathbb{R}^3$ and let r denote the distance to the origin: $r = (x_1^2 + x_2^2 + x_3^2)^{1/2}$; we define the basic weight $\rho(r)$:

$$\rho(r) = \sqrt{1 + r^2}\,.$$

For any multi-index m in $\mathbb{N}^3$, we denote by ∂^m the differential operator of order m:

$$\partial^m = \frac{\partial^{|m|}}{\partial x_1^{m_1} \partial x_2^{m_2} \partial x_3^{m_3}} \quad, \quad \text{with} \quad |m| = m_1 + m_2 + m_3\,.$$

Recall that $\mathcal{D}(\Omega)$ denotes the set of $\mathcal{C}^\infty$ functions with compact support in Ω and $\mathcal{D}'(\Omega)$ denotes its dual space, its elements being called distributions. Then, for all m in $\mathbb{N}$ and all k in $\mathbb{Z}$, we define the weighted space:

$$W_k^m(\Omega) = \left\{ v \in \mathcal{D}'(\Omega)\,;\, \forall \alpha \in \mathbb{N}^3\,,\, 0 \le |\alpha| \le m\,,\, \rho(r)^{|\alpha|-m+k} \partial^\alpha v \in L^2(\Omega) \right\}\,,$$

which is a Hilbert space for the norm:

$$\|v\|_{W_k^m(\Omega)} = \left\{ \sum_{|\alpha|=0}^m \|\rho(r)^{|\alpha|-m+k} \partial^\alpha v\|_{L^2(\Omega)}^2 \right\}^{1/2}\,,$$

where $\|\cdot\|_{L^2(\Omega)}$ denotes the standard norm of $L^2(\Omega)$. We shall denote by $((\cdot,\cdot))_{m,k,\Omega}$ the corresponding scalar product. In addition, we shall sometimes use the following seminorm:

$$|v|_{W_k^m(\Omega)} = \left\{ \sum_{|\alpha|=m} \|\rho(r)^k \partial^\alpha v\|_{L^2(\Omega)}^2 \right\}^{1/2}\,.$$

Note that in this definition, the superscript m denotes the highest order of derivative and the subscript k denotes the exponent of the weight attached to this derivative. In particular

$$W_0^0(\Omega) = L^2(\Omega)\,.$$

The number 1 in the weight $\rho(r)$ is only added so that multiplication by any power of $\rho(r)$ has no influence at the origin; therefore, the functions of $W_k^m(\Omega)$ belong to $H^m(\mathcal{O})$ for all bounded subsets $\mathcal{O}$ of Ω. Moreover, $\rho(r)$ is of the order of r at infinity and for this reason, it is easy to see that the functions of $W_k^m(\Omega)$ are tempered distributions. An important result, established by Hanouzet in [21], for domains with a Lipschitz-continuous boundary is that:

$$\mathcal{D}(\overline{\Omega}) \text{ is dense in } W_k^m(\Omega).$$

35

By virtue of this density and the fact that the functions of $W_k^m(\Omega)$ have the regularity of H^m in the neighbourhood of the boundary Γ, the traces on Γ: $\gamma_0, \gamma_1, \ldots, \gamma_{m-1}$, can be defined as in H^m and the usual trace theorems hold: for any $m \geq 1$, if the boundary Γ is of class $\mathcal{C}^{m-1,1}$, and for each integer i with $0 \leq i \leq m-1$, there exists a constant C such that

$$\forall v \in W_k^m(\Omega), \quad \|\gamma_i v\|_{H^{m-i-\frac{1}{2}}(\Gamma)} \leq C \|v\|_{W_k^m(\Omega)}.$$

With this space, we associate the two following spaces:

$$\overset{\circ}{W}{}_k^m(\Omega) = \text{ the closure of } \mathcal{D}(\Omega) \text{ in } W_k^m(\Omega),$$

$$W_{-k}^{-m}(\Omega) = \text{ the dual space of } \overset{\circ}{W}{}_k^m(\Omega),$$

equipped with the usual dual norm and we denote by $\langle \cdot, \cdot \rangle$ the duality pairing between these two spaces. This extends the definition of the W_k^m spaces to all integers k and m in $\mathbb{Z}$. Furthermore, as in a bounded domain, we have, for $m \geq 1$:

$$\overset{\circ}{W}{}_k^m(\Omega) = \{v \in W_k^m(\Omega); \gamma_i v = 0 \text{ on } \Gamma, \ 0 \leq i \leq m-1\}$$

and the following extension of the Poincaré inequality holds: for each $m \geq 1$ and for all k in $\mathbb{Z}$ there exists a constant C such that

$$\forall v \in \overset{\circ}{W}{}_k^m(\Omega), \quad \|v\|_{W_k^m(\Omega)} \leq C |v|_{W_k^m(\Omega)}, \tag{1.1}$$

i.e. the seminorm $|\cdot|_{W_k^m(\Omega)}$ is a norm on $\overset{\circ}{W}{}_k^m(\Omega)$ equivalent to the norm $\|\cdot\|_{W_k^m(\Omega)}$. Note also that when $m = 0$, the space $\overset{\circ}{W}{}_k^0(\Omega)$ imposes no boundary condition on its functions.

Remark 1.1. It is well–known that, since Ω is unbounded in all directions, the Poincaré inequality (1.1) is false in ordinary Sobolev spaces. It is valid here because the spaces have adequate weights at infinity. $\diamond$

In analogy with the space $H(\text{div})$, we define the following subspace of $W_k^0(\Omega)^3$ for all integers k in $\mathbb{Z}$:

$$H_k(\text{div}; \Omega) = \left\{\vec{v} \in \mathcal{D}'(\Omega)^3 \ ; \ \rho(r)^k \vec{v} \in L^2(\Omega)^3, \rho(r)^{k+1} \text{div}\, \vec{v} \in L^2(\Omega)\right\},$$

equipped with the norm

$$\|\vec{v}\|_{H_k(\text{div};\Omega)} = \left\{\|\rho(r)^k \vec{v}\|_{L^2(\Omega)}^2 + \|\rho(r)^{k+1} \text{div}\, \vec{v}\|_{L^2(\Omega)}^2\right\}^{1/2}.$$

The argument used by Hanouzet to prove the density of $\mathcal{D}(\overline{\Omega})$ in $W_k^m(\Omega)$ can be easily adapted to establish that $\mathcal{D}(\overline{\Omega})^3$ is dense in $H_k(\mathrm{div};\Omega)$. Therefore, denoting by $\vec{n}$ the exterior unit normal to the boundary Γ, the normal trace $\vec{v} \cdot \vec{n}$ can be defined in $H^{-\frac{1}{2}}(\Gamma)$ for functions of $H_k(\mathrm{div};\Omega)$, where $H^{-\frac{1}{2}}(\Gamma)$ denotes the dual space of $H^{\frac{1}{2}}(\Gamma)$. It satisfies the trace theorem: for each integer k in $\mathbb{Z}$, there exists a constant C such that

$$\forall \vec{v} \in H_k(\mathrm{div};\Omega)\ ,\ \|\vec{v}\cdot\vec{n}\|_{H^{-\frac{1}{2}}(\Gamma)} \leq C\|\vec{v}\|_{H_k(\mathrm{div};\Omega)}\ ,$$

and the following Green's formula holds:

$$\forall \vec{v} \in H_k(\mathrm{div};\Omega), \forall \varphi \in W_{-k}^1(\Omega)\ ,\ \langle \vec{v}\cdot\vec{n}, \gamma_0\varphi\rangle_\Gamma = \int_\Omega (\mathrm{div}\,\vec{v})\varphi\,d\mathbf{x} + \int_\Omega \vec{v}\cdot\nabla\varphi\,d\mathbf{x}\,,\quad (1.2)$$

where $\langle\cdot,\cdot\rangle_\Gamma$ denotes the duality product between $H^{-\frac{1}{2}}(\Gamma)$ and $H^{\frac{1}{2}}(\Gamma)$. The closure of $\mathcal{D}(\Omega)^3$ in $H_k(\mathrm{div};\Omega)$ is denoted by $\overset{\circ}{H}_k(\mathrm{div};\Omega)$ and can be characterized by

$$\overset{\circ}{H}_k(\mathrm{div};\Omega) = \{\vec{v}\in H_k(\mathrm{div};\Omega)\,;\,\vec{v}\cdot\vec{n} = 0 \text{ on } \Gamma\}\,.$$

Its dual space is denoted by $H_{-k}^{-1}(\mathrm{div};\Omega)$ and by virtue of the above density, this dual space is a space of distributions.

Remark 1.2. If $\vec{v}$ belongs to $H_k(\mathrm{div};\Omega)$ for any integer $k \geq 1$, then $\mathrm{div}\,\vec{v}$ is in $L^1(\Omega)$ and Green's formula (1.2) yields

$$\langle \vec{v}\cdot\vec{n}, 1\rangle_\Gamma = \int_\Omega \mathrm{div}\,\vec{v}\,d\mathbf{x}\,. \quad (1.3)$$

But when $k \leq 0$, then $\mathrm{div}\,\vec{v}$ is not necessarily in $L^1(\Omega)$ and, in contrast to a bounded domain, (1.3) is generally not valid. In particular, $\mathrm{div}\,\vec{v} = 0$ does not necessarily mean that $\langle \vec{v}\cdot\vec{n}, 1\rangle_\Gamma = 0$. $\qquad\diamond$

Here are some useful properties of the W_k^m spaces.

1. For all integers m and k in $\mathbb{Z}$, all elements of $W_k^m(\Omega)$ have their gradient in $W_k^{m-1}(\Omega)^3$.

2. For all integers m and k in $\mathbb{Z}$,

$$\forall n \in \mathbb{Z}, \text{ with } n \leq m-k-2\,,\quad \mathbb{P}_n \subset W_k^m(\Omega)\,, \quad (1.4)$$

where $\mathbb{P}_n$ denotes the space of all polynomials (of three variables) of degree at most n, with the convention that the space is reduced to zero when n is negative. Thus the difference $m-k$ is an important parameter of the space $W_k^m(\Omega)$.

3. For any integer $m \geq 1$ and for all integers k in $\mathbb{Z}$, the seminorm $|v|_{W_k^m(\Omega)}$ is a norm on the quotient space $W_k^m(\Omega)/\mathbb{P}_n$, equivalent to the quotient norm, where

$$n = \min(m-1, m-k-2),$$

i.e. there exists a constant $C > 0$ such that

$$\forall v \in W_k^m(\Omega)/I\!P_n , \quad \|v\|_{W_k^m(\Omega)/I\!P_n} \le C|v|_{W_k^m(\Omega)} . \tag{1.5}$$

As a consequence of this last property, when n is negative, *i.e.*, when $k > m - 2$, the seminorm $|v|_{W_k^m(\Omega)}$ is a norm equivalent to $\|v\|_{W_k^m(\Omega)}$ on the whole space $W_k^m(\Omega)$. For example, the space $W_0^1(\Omega)$ contains no polynomials and the seminorm $|v|_{W_0^1(\Omega)} = \|\nabla v\|_{L^2(\Omega)}$ is a norm on $W_0^1(\Omega)$ equivalent to the complete norm $\|v\|_{W_0^1(\Omega)}$.

We end this introduction with two theorems, proved by Giroire in [17], concerning isomorphism properties of the Laplace operator in $\mathbb{R}^3$. Beforehand, we introduce the following notation:

$$\mathcal{P}_n^\Delta \text{ is the subspace of all harmonic polynomials of } I\!P_n ,$$

with the convention that $\mathcal{P}_n^\Delta$ is reduced to zero when n is negative.

Theorem 1.3. *For all integers m in $\mathbb{Z}$ and $k \le 0$, the Laplace operator Δ is an isomorphism from the quotient space $W_{m+k}^{m+1}(\mathbb{R}^3)/\mathcal{P}_{-k-1}^\Delta$ onto $W_{m+k}^{m-1}(\mathbb{R}^3)$.*

Theorem 1.4. *For all integers m in $\mathbb{Z}$ and $k \ge 1$, the Laplace operator Δ is an isomorphism from $W_{m+k}^{m+1}(\mathbb{R}^3)$ onto the subspace of $W_{m+k}^{m-1}(\mathbb{R}^3)$ orthogonal to $\mathcal{P}_{k-1}^\Delta$ (for the duality pairing).*

Harmonic functions will play an important role in the sequel. In particular, we shall often use the following properties, that can be found in [17]:
1. The only tempered distributions that are harmonic in $\mathbb{R}^3$ are polynomials. (1.6)
2. For all integers $n \ge 2$, $I\!P_{n-2} = \Delta(I\!P_n)$. (1.7)

2. The divergence and curl in $\mathbb{R}^3$

In this section, we establish important properties of the divergence (and consequently of the gradient) and curl operators in the whole space. They are easier to prove than in exterior domains because they involve no boundary conditions and, as mentioned in the introduction, they are very useful in proving analogous results in exterior domains.

To begin with, we introduce three spaces related to the kernel of the divergence. First, recall the classical space:

$$\mathcal{V}(\mathbb{R}^3) = \left\{\vec{\varphi} \in \mathcal{D}(\mathbb{R}^3)^3 ; \operatorname{div} \vec{\varphi} = 0\right\} ; \tag{2.1}$$

then for all integers m and k in $\mathbb{Z}$, we define

$$V_k^m(\mathbb{R}^3) = \left\{\vec{v} \in W_k^m(\mathbb{R}^3)^3 ; \operatorname{div} \vec{v} = 0\right\} , \tag{2.2}$$

where, if $m \le 0$, the divergence is taken in the sense of distributions in $W_k^{m-1}(\mathbb{R}^3)$, and finally, we define the polar space of $V_k^m(\mathbb{R}^3)$:

$$\left[V_k^m(\mathbb{R}^3)\right]^0 = \left\{\vec{f} \in W_{-k}^{-m}(\mathbb{R}^3)^3 ; \langle \vec{f}, \vec{v} \rangle = 0 \ \forall \vec{v} \in V_k^m(\mathbb{R}^3)\right\} . \tag{2.3}$$

Our first lemma establishes an isomorphism result for the divergence.

Lemma 2.1. *For all integers m in $\mathbb{Z}$ and $k \leq 0$, the divergence is an isomorphism from $W^m_{m+k}(\mathbb{R}^3)^3/V^m_{m+k}(\mathbb{R}^3)$ onto $W^{m-1}_{m+k}(\mathbb{R}^3)$.*

Proof : Take any z in $W^{m-1}_{m+k}(\mathbb{R}^3)$; according to Theorem 1.3, the problem $\Delta\varphi = z$ has a unique solution φ in $W^{m+1}_{m+k}(\mathbb{R}^3)/\mathcal{P}^{\Delta}_{-k-1}$ and the representative $\dot\varphi$ of φ defined by:

$$\forall q \in \mathcal{P}^{\Delta}_{-k-1} \qquad ((\dot\varphi, q))_{m+1, m+k, \mathbb{R}^3} = 0 \tag{2.4}$$

satisfies

$$\|\dot\varphi\|_{W^{m+1}_{m+k}(\mathbb{R}^3)} \leq C\|z\|_{W^{m-1}_{m+k}(\mathbb{R}^3)}.$$

Let us take $\vec{v} = \nabla\dot\varphi$. Then $\vec{v}$ belongs to $W^m_{m+k}(\mathbb{R}^3)^3$, div $\vec{v} = z$ and

$$\|\vec{v}\|_{W^m_{m+k}(\mathbb{R}^3)} \leq C\|z\|_{W^{m-1}_{m+k}(\mathbb{R}^3)}.$$

The mapping $z \mapsto \vec{v}$ is linear, the function $\vec{v}$ is obviously unique in the quotient space $W^m_{m+k}(\mathbb{R}^3)^3/V^m_{m+k}(\mathbb{R}^3)$ and the quotient norm is of course not greater than the norm of the representative we have constructed.

Thus we have established that the divergence operator is an isomorphism from the quotient space $W^m_{m+k}(\mathbb{R}^3)^3/V^m_{m+k}(\mathbb{R}^3)$ onto $W^{m-1}_{m+k}(\mathbb{R}^3)$. $\diamondsuit$

The crucial step in the above proof is the solution of the Laplace equation. For nonpositive values of k, the solution exists without orthogonality restriction on the right-hand side. When $k = 1$, the range space of the divergence is necessarily orthogonal to constants, so that the above proof can be applied to show that the divergence is an isomorphism from $W^m_{m+1}(\mathbb{R}^3)^3$ onto $W^{m-1}_{m+1}(\mathbb{R}^3)\perp\mathbb{R}$. But, when $k \geq 2$, because of the orthogonality condition in the statement of Theorem 1.4, this proof no longer applies to the space $W^{m-1}_{m+k}(\mathbb{R}^3)$; more precisely, in such spaces, the divergence cannot be lifted by means of a gradient. But of course, this is not necessary and this difficulty will be overcome by a dual argument lifting the gradient instead of the divergence. This is the object of the next proposition.

Proposition 2.2. *Let the integers m and k belong to $\mathbb{Z}$ with $k \leq -1$. Then each distribution q of $\mathcal{D}'(\mathbb{R}^3)$ such that ∇q is in $W^m_{m+k}(\mathbb{R}^3)^3$ belongs itself to $W^{m+1}_{m+k}(\mathbb{R}^3)$ and there exists a constant C, independent of q, such that:*

$$\|q\|_{W^{m+1}_{m+k}(\mathbb{R}^3)/\mathbb{R}} \leq C\|\nabla q\|_{W^m_{m+k}(\mathbb{R}^3)}. \tag{2.5}$$

Proof : The idea is to construct an element of $W^{m+1}_{m+k}(\mathbb{R}^3)$ with the same gradient as q. Set $f = \text{div}\,(\nabla q)$. As f belongs to $W^{m-1}_{m+k}(\mathbb{R}^3)$, it follows from Theorem 1.3 that there exists a unique φ in $W^{m+1}_{m+k}(\mathbb{R}^3)/\mathcal{P}^{\Delta}_{-k-1}$ such that $\Delta\varphi = f$ and, as above, the representative $\dot\varphi$ of φ defined by (2.4) satisfies $\|\dot\varphi\|_{W^{m+1}_{m+k}(\mathbb{R}^3)} \leq C_1\|\nabla q\|_{W^m_{m+k}(\mathbb{R}^3)}$ with a constant C_1 independent of q.

Then $\nabla q - \nabla\dot\varphi$ is a vector of $W^m_{m+k}(\mathbb{R}^3)^3$ with zero divergence and zero **curl**. Using (1.6) and (1.4), it is easy to prove that each of its components belongs to $\mathcal{P}^\Delta_{-k-2}$. Hence since $\nabla q - \nabla\dot\varphi$ is a gradient, there exists a polynomial p in $\mathbb{P}_{-k-1}$ such that $\nabla q - \nabla\dot\varphi = \nabla p$. Set $u = \dot\varphi + p$; u belongs to $W^{m+1}_{m+k}(\mathbb{R}^3)$ and

$$\nabla q = \nabla u = \nabla(\dot\varphi + p).$$

As a consequence, q and u differ by a constant (this result, valid for two distributions, is not altogether trivial and can be found in Schwarz [30] and Simon [31]). Therefore, q also belongs to $W^{m+1}_{m+k}(\mathbb{R}^3)$ and

$$\|q\|_{W^{m+1}_{m+k}(\mathbb{R}^3)/\mathbb{R}} = \|u\|_{W^{m+1}_{m+k}(\mathbb{R}^3)/\mathbb{R}}.$$

When m is nonnegative, (2.5) follows directly from the equivalence of norms (1.5). Otherwise, observe that, since $\mathbb{P}_{-k-1}$ is a finite-dimensional subspace of $W^{m+1}_{m+k}(\mathbb{R}^3)$, there exists a constant C_2 such that

$$\|p\|_{W^{m+1}_{m+k}(\mathbb{R}^3)/\mathbb{R}} \le C_2\|\nabla p\|_{W^m_{m+k}(\mathbb{R}^3)}$$
$$\le C_2\left\{\|\nabla q\|_{W^m_{m+k}(\mathbb{R}^3)} + \|\nabla\dot\varphi\|_{W^m_{m+k}(\mathbb{R}^3)}\right\}$$
$$\le C_3\|\nabla q\|_{W^m_{m+k}(\mathbb{R}^3)}.$$

Therefore,

$$\|q\|_{W^{m+1}_{m+k}(\mathbb{R}^3)/\mathbb{R}} \le \left\{\|\dot\varphi\|_{W^{m+1}_{m+k}(\mathbb{R}^3)} + \|p\|_{W^{m+1}_{m+k}(\mathbb{R}^3)/\mathbb{R}}\right\} \le C_4\|\nabla q\|_{W^m_{m+k}(\mathbb{R}^3)}.$$

$\diamond$

By combining Lemma 2.1 and Proposition 2.2, we can extend Lemma 2.1 to positive values of k.

Theorem 2.3. (The "inf-sup" condition in $\mathbb{R}^3$) *For all integers m and k in $\mathbb{Z}$, the following equivalent properties hold:*
1. *The divergence is an isomorphism from $W^m_{m+k}(\mathbb{R}^3)^3/V^m_{m+k}(\mathbb{R}^3)$ onto $W^{m-1}_{m+k}(\mathbb{R}^3)$ (or $W^{m-1}_{m+k}(\mathbb{R}^3)\bot\mathbb{R}$ when $k \ge 1$).*
2. *The gradient is an isomorphism from $W^{1-m}_{-(m+k)}(\mathbb{R}^3)$ (or $W^{1-m}_{-(m+k)}(\mathbb{R}^3)/\mathbb{R}$ when $k \ge 1$) onto $\left[V^m_{m+k}(\mathbb{R}^3)\right]^0$.*

Proof : The statements of parts 1 and 2 are dual and therefore equivalent propositions. Lemma 2.1 has proved part 1 for negative k; therefore, it suffices either to establish part 1 or part 2 for strictly positive k. Let us prove the latter. It follows from Proposition 2.2, that the gradient operator is an isomorphism from

$W^{1-m}_{-(m+k)}(\mathbb{R}^3)/\mathbb{R}$ onto its range. Hence, this range is a closed subspace of $W^{-m}_{-(m+k)}(\mathbb{R}^3)^3$ and the Closed Range Theorem of Banach implies that

$$\text{Range}(\nabla) = [\text{Ker}(-\text{div})]^0 = [V^m_{m+k}(\mathbb{R}^3)]^0 .$$

This is precisely the statement of part 2. ◇

Proposition 2.2 characterizes distributions by means of their gradients, for negative values of k. This is a very important result, but because of the orthogonality restriction of Theorem 1.4, its proof does not extend to all positive values of k. We shall prove it further on via a density argument based on the properties of the curl operator.

Now we turn to the curl operator. For all integers $k \geq 0$, we define the following subspace of $(\mathcal{P}^\Delta_k)^3$:

$$\mathcal{G}_k = \{\nabla q \, ; \, q \in \mathcal{P}^\Delta_{k+1}\} . \tag{2.6}$$

The following preliminary lemma shows how to construct a polynomial vector with prescribed curl and divergence.

Lemma 2.4. *Let $k \geq 1$ be an integer. Let $\vec{q}$ be a given polynomial vector in $\mathbb{P}^3_{k-1}$ such that $\text{div}\,\vec{q} = 0$ and let r be a given polynomial in $\mathbb{P}_{k-1}$. Then there exists a polynomial vector $\vec{s}$ unique in $\mathbb{P}^3_k/\mathcal{G}_k$ such that*

$$\mathbf{curl}\,\vec{s} = \vec{q} \qquad and \qquad \text{div}\,\vec{s} = r .$$

The mapping $(\vec{q}, r) \mapsto \vec{s}$ is linear.

Proof : Set $\vec{s} = (s_1, s_2, s_3)$, $\vec{q} = (q_1, q_2, q_3)$ and exceptionally, let ∂_i denote the partial derivative $\partial/\partial x_i$ and Υ_i denote the primitive that vanishes when $x_i = 0$, *i.e.*

$$\Upsilon_i(x_i^{\alpha_i} x_j^{\alpha_j} x_k^{\alpha_k}) = \frac{1}{\alpha_i + 1} x_i^{\alpha_i+1} x_j^{\alpha_j} x_k^{\alpha_k} .$$

On a polynomial space, Υ_i is well defined, commutes with ∂_j and $\Upsilon_i \partial_i$ is the identity mapping.

To begin with, let us construct a polynomial vector $\vec{s}$ in $\mathbb{P}^3_k$ that satisfies the first condition of the lemma, *i.e.* such that

$$\partial_2 s_3 - \partial_3 s_2 = q_1 \quad , \quad \partial_3 s_1 - \partial_1 s_3 = q_2 \quad , \quad \partial_1 s_2 - \partial_2 s_1 = q_3 .$$

Take for instance $s_3 = 0$; to satisfy the first two equations we choose

$$s_2 = -\Upsilon_3 q_1 \qquad and \qquad s_1 = \Upsilon_3 q_2 ,$$

which are both polynomials of $\mathbb{P}_k$. Since by assumption,

$$\partial_1 q_1 + \partial_2 q_2 + \partial_3 q_3 = 0 ,$$

41

it is easy to check that the third equation is necessarily satisfied.

Once the first condition of the lemma is fulfilled, the second condition follows readily from property (1.7). Indeed, div $\vec{s} - r$ belongs to $\mathbb{P}_{k-1}$ and according to (1.7), there exists p in $\mathbb{P}_{k+1}$ such that

$$\Delta p = \operatorname{div} \vec{s} - r\,.$$

Then $\vec{s} - \nabla p$ satisfies both conditions of the lemma.

As far as uniqueness is concerned, if a polynomial $\vec{s}$ of $\mathbb{P}_k^3$ satisfies

$$\mathbf{curl}\,\vec{s} = \vec{0} \qquad \text{and} \qquad \operatorname{div} \vec{s} = 0\,,$$

then $\vec{s} = \nabla p$ for some p in $\mathbb{P}_{k+1}$ and $\Delta p = 0$. Hence, $\vec{s}$ belongs to $\mathcal{G}_k$. The linearity of the mapping $(\vec{q}, r) \mapsto \vec{s}$ follows from this uniqueness. $\diamond$

The following theorem characterizes the vector potential of a divergence-free vector field.

Theorem 2.5. *Let m and k belong to $\mathbb{Z}$ with $k \leq 2$. Let $\vec{u}$ be a vector field in $W_{m+k}^m(\mathbb{R}^3)^3$ such that*
$$\operatorname{div} \vec{u} = 0\,.$$

Then $\vec{u}$ has a unique vector potential $\vec{\varphi}$ in $W_{m+k}^{m+1}(\mathbb{R}^3)^3/\mathcal{G}_{-k-1}$ such that

$$\vec{u} = \mathbf{curl}\,\vec{\varphi} \qquad , \qquad \operatorname{div} \vec{\varphi} = 0\,, \tag{2.7}$$

$$\|\vec{\varphi}\|_{W_{m+k}^{m+1}(\mathbb{R}^3)^3/\mathcal{G}_{-k-1}} \leq C\|\vec{u}\|_{W_{m+k}^m(\mathbb{R}^3)}\,. \tag{2.8}$$

When, $k \geq 0$, the vector potential is unique in $W_{m+k}^{m+1}(\mathbb{R}^3)^3$ and (2.8) can be slightly refined:

$$\|\vec{\varphi}\|_{W_{m+k}^{m+1}(\mathbb{R}^3)} \leq C\|\mathbf{curl}\,\vec{u}\|_{W_{m+k}^{m-1}(\mathbb{R}^3)}\,. \tag{2.9}$$

Proof : First note that if $\vec{u}$ has two vector potentials satisfying the statement of the theorem then their difference, say $\vec{\lambda}$, satisfies :

$$\mathbf{curl}\,\vec{\lambda} = \vec{0} \qquad \text{and} \qquad \operatorname{div} \vec{\lambda} = 0\,. \tag{2.10}$$

Therefore, $\vec{\lambda}$ belongs to $\left(\mathcal{P}_{-k-1}^{\Delta}\right)^3$ and more precisely, $\vec{\lambda}$ belongs to $\mathcal{G}_{-k-1}$. This establishes the uniqueness of the vector potential up to a polynomial vector of $\mathcal{G}_{-k-1}$. In particular, when $k \geq 0$, $\mathcal{G}_{-k-1}$ contains only the zero polynomial vector and the vector potential is unique.

To show the existence, let us solve the problem

$$-\Delta \vec{\varphi} = \mathbf{curl}\,\vec{u}\,. \tag{2.11}$$

Assume for the moment that $k \leq 0$. Since $\mathbf{curl}\,\vec{u}$ belongs to $W_{m+k}^{m-1}(\mathbb{R}^3)^3$, it follows from Theorem 1.3 that this problem has a unique solution $\vec{\varphi}$ in $\left[W_{m+k}^{m+1}(\mathbb{R}^3)/\mathcal{P}_{-k-1}^{\Delta}\right]^3$ and the representative $\vec{\varphi}$ defined by (2.4) satisfies

$$\|\vec{\varphi}\|_{W_{m+k}^{m+1}(\mathbb{R}^3)} \leq C\|\mathbf{curl}\,\vec{u}\|_{W_{m+k}^{m-1}(\mathbb{R}^3)}\,.$$

Now, let us show that $\mathrm{div}\,\vec{\varphi}$ belongs to $\mathcal{P}_{-k-2}^{\Delta}$ and $\mathbf{curl}\,\vec{\varphi}-\vec{u}$ belongs to $\left(\mathcal{P}_{-k-2}^{\Delta}\right)^3$. The first property is obtained by taking the divergence of both sides of equation (2.11) and using (1.6) and (1.4). Similarly, the second property is obtained by noting that $\Delta(\mathbf{curl}\,\vec{\varphi}-\vec{u}) = \vec{0}$. Next, observe that $\mathrm{div}(\mathbf{curl}\,\vec{\varphi}-\vec{u}) = 0$. Assume first that $k \leq -2$. It stems from the construction of Lemma 2.4 that we can find a polynomial vector $\vec{p}$ in $I\!P_{-k-1}^3$ such that

$$\mathbf{curl}(\vec{\varphi}-\vec{p}) = \vec{u} \qquad \text{and} \qquad \mathrm{div}(\vec{\varphi}-\vec{p}) = 0\,.$$

As $I\!P_{-k-1}$ is contained in $W_{m+k}^{m+1}(\mathbb{R}^3)$, it is easy to see that $\vec{\varphi}-\vec{p}$ satisfies the statement of the theorem. When $k = 0$ or -1, the above result holds with $\vec{p} = \vec{0}$. Indeed, $\mathrm{div}\,\vec{\varphi} = 0$ and $\mathbf{curl}\,\vec{\varphi} = \vec{u}$.

Finally, consider the case where $k = 1$ or 2. It follows from Theorem 1.4 that problem (2.11) has a (unique) solution $\vec{\varphi}$ in $W_{m+k}^{m+1}(\mathbb{R}^3)^3$, provided $\mathbf{curl}\,\vec{u}$ is orthogonal to $\mathcal{P}_0^{\Delta}$ (when $k = 1$) or $\mathcal{P}_1^{\Delta}$ (when $k = 2$). The first orthogonality is obvious, and the second orthogonality stems from the fact that $\vec{u}$ is divergence-free. Indeed,

$$\langle \vec{u}, \mathbf{curl}\,\vec{q}\rangle = 0$$

if there exists μ in $I\!P_1$ such that $\mathbf{curl}\,\vec{q} = \nabla\mu$. But this is always the case when $\vec{q}$ belongs to $I\!P_1^3$. $\diamond$

Remark 2.6. Because of the orthogonality condition in Theorem 1.4, we cannot extend Theorem 2.5 to arbitrary positive integers. Nevertheless, proceeding by duality, we can still construct a vector potential in this case, but it will not necessarily be divergence-free. However, this result which is fairly technical will not prove to be very useful in exterior domains and therefore we shall skip it. The reader can find it in [12]. $\diamond$

An important application of Theorem 2.5 is the following density result.

Theorem 2.7. The density of smooth divergence-free functions. *For all integers m and k in $\mathbb{Z}$ with $k \leq 2$, the space $\mathcal{V}(\mathbb{R}^3)$ is dense in $V_{m+k}^{m}(\mathbb{R}^3)$.*

Proof : Let $\vec{v}$ belong to $V_{m+k}^{m}(\mathbb{R}^3)$; it follows from the construction of Theorem 2.5 that, by choosing an adequate representative, there exists $\vec{\varphi}$ in $W_{m+k}^{m+1}(\mathbb{R}^3)^3$ such that

$$\vec{v} = \mathbf{curl}\,\vec{\varphi} \qquad \text{and} \qquad \mathrm{div}\,\vec{\varphi} = 0\,,$$

(although this last property is irrelevant in this proof). Now, since $\mathcal{D}(\mathbb{R}^3)$ is dense in $W^{m+1}_{m+k}(\mathbb{R}^3)$, there exists a sequence $\vec{\varphi}_\ell$ in $\mathcal{D}(\mathbb{R}^3)^3$ such that

$$\lim_{\ell \to \infty} \vec{\varphi}_\ell = \vec{\varphi} \qquad \text{in } W^{m+1}_{m+k}(\mathbb{R}^3)^3 \,.$$

Hence the sequence $\vec{v}_\ell = \mathbf{curl}\, \vec{\varphi}_\ell$ belongs to $\mathcal{V}(\mathbb{R}^3)$ and

$$\lim_{\ell \to \infty} \vec{v}_\ell = \lim_{\ell \to \infty} \mathbf{curl}\, \vec{\varphi}_\ell = \mathbf{curl}\, \vec{\varphi} = \vec{v} \qquad \text{in } W^m_{m+k}(\mathbb{R}^3)^3 \,.$$

This proves the theorem. $\diamond$

Remark 2.8. In fact, by using Theorem 2.3, a duality argument and de Rham's Theorem that says that a distribution that vanishes on $\mathcal{V}$ is necessarily a gradient (*cf.* de Rham [29], Tartar [32] or Simon [31]), the statement of Theorem 2.7 extends easily to all positive values of k. $\diamond$

With Theorem 2.7, we can complete the statement of Proposition 2.2.

Corollary 2.9. *Let m belong to $\mathbb{Z}$ and k belong to $\mathbb{N}$ and let q be a distribution of $\mathcal{D}'(\mathbb{R}^3)$ such that ∇q belongs to $W^m_{m+k}(\mathbb{R}^3)^3$. Then q has the unique decomposition*

$$q = s + c \,,$$

where c is a constant and s belongs to $W^{m+1}_{m+k}(\mathbb{R}^3)$. Moreover,

$$\|s\|_{W^{m+1}_{m+k}(\mathbb{R}^3)} \leq C \|\nabla q\|_{W^m_{m+k}(\mathbb{R}^3)} \,.$$

Proof : Set $\vec{\ell} = \nabla q$ and let $\vec{\varphi}$ belong to $\mathcal{V}(\mathbb{R}^3)$. Then

$$\langle \vec{\ell}, \vec{\varphi} \rangle = \langle \nabla q, \vec{\varphi} \rangle = -\langle q, \operatorname{div} \vec{\varphi} \rangle = 0 \,.$$

Therefore, $\vec{\ell}$ vanishes on $\mathcal{V}(\mathbb{R}^3)$ and since $\mathcal{V}(\mathbb{R}^3)$ is dense in $V^{-m}_{-m-k}(\mathbb{R}^3)$, this implies that $\vec{\ell}$ belongs to $\left[V^{-m}_{-m-k}(\mathbb{R}^3) \right]^0$. Hence, Theorem 2.3 implies that there exists a unique s in $W^{m+1}_{m+k}(\mathbb{R}^3)$ such that $\vec{\ell} = \nabla s$ and

$$\|s\|_{W^{m+1}_{m+k}(\mathbb{R}^3)} \leq C \|\vec{\ell}\|_{W^m_{m+k}(\mathbb{R}^3)} \,.$$

As s and q have the same gradient, they differ by a constant. The uniqueness of the decomposition follows from the fact that the constant functions do not belong to $W^{m+1}_{m+k}(\mathbb{R}^3)$ for $k \geq 0$. $\diamond$

3. The divergence in exterior domains

In this section, we propose to show that the divergence operator is an isomorphism between adequate spaces. This result is not only crucial for the Stokes problem but it also permits to characterize distributions by means of their gradients.

Let us extend to exterior domains the spaces defined in (2.1), (2.2) and (2.3):

$$\mathcal{V}(\Omega) = \left\{ \vec{\varphi} \in \mathcal{D}(\Omega)^3 \; ; \; \operatorname{div} \vec{\varphi} = 0 \right\} , \qquad (3.1)$$

and for all integers $m \geq 1$ and k in $\mathbb{Z}$,

$$V_k^m(\Omega) = \left\{ \vec{v} \in \overset{\circ}{W}{}_k^m(\Omega)^3 \; ; \; \operatorname{div} \vec{v} = 0 \right\} . \qquad (3.2a)$$

This definition does not extend satisfactorily to $m = 0$ because it imposes no boundary condition on the functions. We replace it by

$$V_k^0(\Omega) = \left\{ \vec{v} \in \overset{\circ}{H}_k(\operatorname{div}; \Omega) \; ; \; \operatorname{div} \vec{v} = 0 \right\} . \qquad (3.2b)$$

This definition can also be extended to negative m, but we shall not need it here. With these spaces, we associate the polar space of $V_k^m(\Omega)$

$$[V_k^m(\Omega)]^0 = \left\{ \vec{f} \in W_{-k}^{-m}(\Omega)^3 \; ; \; \langle \vec{f}, \vec{v} \rangle = 0 \; \forall \vec{v} \in V_k^m(\Omega) \right\} \quad \text{for } m \geq 1, \qquad (3.3a)$$

$$[V_k^0(\Omega)]^0 = \left\{ \vec{f} \in H_{-k}^{-1}(\operatorname{div}; \Omega) \; ; \; \langle \vec{f}, \vec{v} \rangle = 0 \; \forall \vec{v} \in V_k^0(\Omega) \right\} \quad \text{for } m = 0. \qquad (3.3b)$$

Our first lemma states an isomorphism on the divergence.

Lemma 3.1. *Let the domain Ω' be Lipschitz-continuous. For all integers $m \geq 1$ and all integers k in $\mathbb{Z}$, the divergence operator is an isomorphism from $\overset{\circ}{W}{}_{m+k}^m(\Omega)^3 / V_{m+k}^m(\Omega)$ onto $\overset{\circ}{W}{}_{m+k}^{m-1}(\Omega)$ if $k \leq 0$ or onto $\overset{\circ}{W}{}_{m+k}^{m-1}(\Omega) \perp \mathbb{R}$ if $k \geq 1$.*

Proof : Let z belong to $\overset{\circ}{W}{}_{m+k}^{m-1}(\Omega)$ if $k \leq 0$ or $\overset{\circ}{W}{}_{m+k}^{m-1}(\Omega) \perp \mathbb{R}$ if $k \geq 1$; we want to find $\vec{v}$ in $\overset{\circ}{W}{}_{m+k}^m(\Omega)^3$ such that $\operatorname{div} \vec{v} = z$. First observe that when $k \geq 1$, if $\vec{v}$ exists, it belongs at least to $H_1(\operatorname{div}; \Omega)$ and by applying (1.3), we have

$$\int_\Omega \operatorname{div} \vec{v} \, d\mathbf{x} = 0 ,$$

since $\vec{v}$ vanishes on Γ. Hence, for such a function $\vec{v}$ to exist, z must necessarily be orthogonal to constants. Note also that this condition is not necessary when $k \leq 0$.

Next, let us extend z by zero in Ω'. Owing to the boundary conditions on z, the extended function that we still denote by z belongs to $W_{m+k}^{m-1}(\mathbb{R}^3)$ if $k \leq 0$ or it belongs to $W_{m+k}^{m-1}(\mathbb{R}^3) \perp \mathbb{R}$ if $k \geq 1$. Therefore, owing to Theorem 2.3, there exists a unique $\vec{\varphi}$ in $W_{m+k}^m(\mathbb{R}^3)^3$, orthogonal to $V_{m+k}^m(\mathbb{R}^3)$, such that

$$\operatorname{div} \vec{\varphi} = z \quad \text{in } \mathbb{R}^3,$$

$$\|\vec{\varphi}\|_{W_{m+k}^m(\mathbb{R}^3)} \leq C_1 \|z\|_{W_{m+k}^{m-1}(\Omega)},$$

where C_1 is a constant that depends only on m and k.

It remains to find a divergence-free function that has the same traces as $\vec{\varphi}$ on Γ. Assume first that Ω' is of class $\mathcal{C}^{m-1,1}$. As Ω' is bounded, we can fix once and for all a ball B_{R_0}, centered at the origin and with radius R_0, such that $\overline{\Omega'} \subset B_{R_0}$. Then we set

$$\Omega_{R_0} = \Omega \cap B_{R_0}.$$

Since Ω' is of class $\mathcal{C}^{m-1,1}$, then so is Ω_{R_0} and therefore $\vec{\varphi}$ has traces up to the order $m-1$ on the boundary of Ω_{R_0}. On the other hand, $\operatorname{div} \vec{\varphi} = z = 0$ in Ω'. Therefore, the traces $\gamma_0 \vec{\varphi}, \gamma_1 \vec{\varphi}, \dots, \gamma_{m-1} \vec{\varphi}$ on Γ satisfy the compatibility conditions of Héron [22] for the existence of a divergence-free lift in Ω_{R_0} (cf. also Amrouche & Girault [2]). Thus we can fix a function $\vec{w}$ in $H^m(\Omega_{R_0})^3$ such that

$$\operatorname{div} \vec{w} = 0 \quad \text{in } \Omega_{R_0},$$

$$\gamma_i \vec{w} = \gamma_i \vec{\varphi} \quad \text{on } \Gamma, \quad \text{for } 0 \leq i \leq m-1,$$

$$\gamma_i \vec{w} = \vec{0} \quad \text{on } \partial B_{R_0}, \quad \text{for } 0 \leq i \leq m-1,$$

and the mapping $\vec{\varphi} \mapsto \vec{w}$ is linear and continuous:

$$\|\vec{w}\|_{H^m(\Omega_{R_0})} \leq C_2 \|\vec{\varphi}\|_{W_{m+k}^m(\mathbb{R}^3)},$$

with a constant C_2 that depends only on Ω_{R_0}, m and k. The function $\vec{w}$ can be extended by zero outside B_{R_0} and owing to its boundary conditions on ∂B_{R_0}, the extended function, still denoted by $\vec{w}$ belongs to $W_{m+k}^m(\Omega)^3$, for any k since its support is bounded. Then the function

$$\vec{v} = \vec{\varphi} - \vec{w}$$

belongs to $\overset{\circ}{W}{}_{m+k}^m(\Omega)^3$ and satisfies

$$\operatorname{div} \vec{v} = z,$$

$$\|\vec{v}\|_{W_{m+k}^m(\Omega)^3 / V_{m+k}^m(\Omega)} \leq \mathcal{B} \|z\|_{W_{m+k}^{m-1}(\Omega)}. \tag{3.4}$$

The mapping $z \mapsto \vec{v}$ is obviously linear.

Finally, we eliminate the regularity assumption on the boundary by taking into account that, in the bounded case, a similar result holds with no restriction on the boundary. Then the desired result is proved by means of a partition of unity. We skip the proof for the sake of conciseness; it can be found in [13]. $\qquad \diamondsuit$

With this lemma, we can establish an "inf-sup" condition.

Theorem 3.2. *Assume that Ω' is Lipschitz-continuous. For all integers $m \geq 1$ and all integers k in $\mathbb{Z}$, the following equivalent properties hold:*

1. *The divergence operator is an isomorphism from $\overset{\circ}{W}{}^{\,m}_{m+k}(\Omega)^3 / V^m_{m+k}(\Omega)$ onto $\overset{\circ}{W}{}^{\,m-1}_{m+k}(\Omega)$ if $k \leq 0$ or onto $\overset{\circ}{W}{}^{\,m-1}_{m+k}(\Omega)\perp\mathbb{R}$ if $k \geq 1$.*

2. *The gradient operator is an isomorphism from $W^{1-m}_{-(m+k)}(\Omega)$ if $k \leq 0$ or $W^{1-m}_{-(m+k)}(\Omega)/\mathbb{R}$ if $k \geq 1$ onto $\left[V^m_{m+k}(\Omega)\right]^0$.*

3. *There exists a constant $\beta > 0$ (in fact $\beta = \frac{1}{\mathcal{B}}$, where $\mathcal{B}$ is the constant of (3.4)) such that:*

$$\inf_{q \in W^{1-m}_{-(m+k)}(\Omega)} \sup_{\vec{v} \in \overset{\circ}{W}{}^{\,m}_{m+k}(\Omega)^3} \frac{\langle q, div\,\vec{v}\rangle}{\|q\|_{W^{1-m}_{-(m+k)}(\Omega)} \|\vec{v}\|_{W^m_{m+k}(\Omega)}} \geq \beta, \qquad (3.5)$$

with the convention that q must be taken in $W^{1-m}_{-(m+k)}(\Omega)/\mathbb{R}$ and its norm replaced by the quotient norm when $k \geq 1$.

Proof: The first part of the theorem is simply Lemma 3.1 and the other two propositions are equivalent to this statement thanks to the classical "inf-sup" condition established by Babuška and Brezzi (*cf.* Babuška [3], Brezzi [5] or Girault & Raviart [14]) in an abstract situation. The correspondence is as follows:

$$X = \overset{\circ}{W}{}^{\,m}_{m+k}(\Omega)^3, \quad X' = W^{-m}_{-(m+k)}(\Omega)^3,$$

$$M = W^{1-m}_{-(m+k)}(\Omega) \text{ if } k \leq 0 \text{ or } W^{1-m}_{-(m+k)}(\Omega)/\mathbb{R} \text{ if } k \geq 1,$$

$$M' = \overset{\circ}{W}{}^{\,m-1}_{m+k}(\Omega) \text{ if } k \leq 0 \text{ or } \overset{\circ}{W}{}^{\,m-1}_{m+k}(\Omega)\perp\mathbb{R} \text{ if } k \geq 1,$$

$$B = div : X \mapsto M', \quad B' = -\nabla : M \mapsto X',$$

$$V = \text{Ker}(div) = V^m_{m+k}(\Omega).$$

The operators div and $-\nabla$ are dual operators in these spaces by virtue of the generalized Green's formula

$$\forall \vec{v} \in \overset{\circ}{W}{}^{\,m}_{m+k}(\Omega)^3, \ \forall q \in W^{1-m}_{-(m+k)}(\Omega), \ \langle div\,\vec{v}, q\rangle = -\langle \vec{v}, \nabla q\rangle. \qquad \diamond$$

Theorem 3.2 yields the following extension of (1.5) for $m \leq -1$:

$$\forall q \in W^{m+1}_{m+k}(\Omega) \quad \|q\|_{W^{m+1}_{m+k}(\Omega)/\mathbb{R}} \leq \mathcal{B}\|\nabla q\|_{W^m_{m+k}(\Omega)} \quad \text{if } k \leq -1, \qquad (3.6a)$$

$$\forall q \in W^{m+1}_{m+k}(\Omega) \quad \|q\|_{W^{m+1}_{m+k}(\Omega)} \leq \mathcal{B}\|\nabla q\|_{W^m_{m+k}(\Omega)} \quad \text{if } k \geq 0, \qquad (3.6b)$$

where $\mathcal{B}$ is the constant of (3.4) with parameters $-m$ and $-k$.

Next, we prove a preliminary lemma that gives a first construction of a vector potential.

Lemma 3.3. *Let Ω' be Lipschitz-continuous. For any integers $m \geq 0$ and k in $\mathbb{Z}$, each function $\vec{v}$ in $V^m_{m+k}(\Omega)$ has a vector potential $\vec{\varphi}$ in $W^{m+1}_{m+k}(\Omega)^3$:*

$$\vec{v} = \mathbf{curl}\, \vec{\varphi}\,.$$

Proof : Let us extend $\vec{v}$ by zero in Ω'. Owing to the boundary conditions imposed by $V^m_{m+k}(\Omega)$, the extended function, that we still denote by $\vec{v}$, belongs to $V^m_{m+k}(\mathbb{R}^3)$. When $k \leq 2$, Theorem 2.5 proves that there exists $\vec{\varphi}$ in $V^{m+1}_{m+k}(\mathbb{R}^3)$ such that

$$\vec{v} = \mathbf{curl}\, \vec{\varphi} \quad \text{in } \mathbb{R}^3\,.$$

When $k \geq 3$, Remark 2.6 states that such $\vec{\varphi}$ exists in $W^{m+1}_{m+k}(\mathbb{R}^3)^3$. $\qquad\qquad \Diamond$

Although the statement of this lemma is crude, since it gives no information about uniqueness and continuity of the vector potential, it permits to prove a useful density result.

Theorem 3.4. *Let Ω' be Lipschitz-continuous. For every integers $m \geq 0$ and k in $\mathbb{Z}$, the space $\mathcal{V}(\Omega)$ is dense in $V^m_{m+k}(\Omega)$.*

Proof : Let $\vec{v}$ belong to $V^m_{m+k}(\Omega)$; then according to Lemma 3.3,

$$\vec{v} = \mathbf{curl}\, \vec{\varphi} \quad \text{in } \Omega\,,$$

for some $\vec{\varphi}$ in $W^{m+1}_{m+k}(\mathbb{R}^3)^3$. The proof, consists in approximating $\vec{\varphi}$ by a sequence of functions with bounded support. For any real number R, let B_R be the ball centered at the origin with radius R. Let a be the standard cut-off function :

$$a \in \mathcal{D}(\mathbb{R}^3) \quad,\quad 0 \leq a(\mathbf{x}) \leq 1 \quad \forall \mathbf{x} \in \mathbb{R}^3 \quad,\quad a \equiv 1 \text{ in } B_1 \quad \text{and} \quad \text{supp}(a) \subset B_2\,.$$

Set

$$\forall \mathbf{x} \in \mathbb{R}^3 \,,\, a_R(\mathbf{x}) = a(\mathbf{x}/R)$$

and take $\vec{\varphi}_R = a_R \vec{\varphi}$. Hanouzet proves in [21] that

$$\lim_{R \to \infty} \vec{\varphi}_R = \vec{\varphi} \quad \text{in } W^{m+1}_{m+k}(\mathbb{R}^3)^3\,.$$

Consider the function $\vec{v}_R = \mathbf{curl}\,(\vec{\varphi}_R)$. It coincides with $\vec{v}$ in a neighbourhood of Γ for large enough R and is obviously divergence-free. Therefore, $\vec{v}_R$ belongs to $V^m_{m+k}(\Omega)$, has a bounded support and

$$\lim_{R \to \infty} \vec{v}_R = \lim_{R \to \infty} curl\,(\vec{\varphi}_R) = \mathbf{curl}\, \vec{\varphi} = \vec{v} \quad \text{in } W^m_{m+k}(\Omega)^3\,.$$

In other words, the set of functions of $V^m_{m+k}(\Omega)$ with bounded support is dense in $V^m_{m+k}(\Omega)$. Then the theorem follows from the fact that in a bounded Lipschitz-continuous domain $\mathcal{O}$, the space $\mathcal{V}(\mathcal{O})$ is dense in $V^m_l(\mathcal{O})$ (for any weight l, as the weight is not significant in a bounded domain). $\qquad\qquad \Diamond$

The next theorem characterizes distributions by means of their gradients. It is partly an application of Theorems 3.2 and 3.4.

Theorem 3.5. *Let Ω' be Lipschitz-continuous and let m and k be two integers in $\mathbb{Z}$. Let q be a distribution of $\mathcal{D}'(\Omega)$ such that ∇q belongs to $W_{m+k}^m(\Omega)^3$. Then, if $k \leq -1$, q belongs to $W_{m+k}^{m+1}(\Omega)$ and*

$$\|q\|_{W_{m+k}^{m+1}(\Omega)/\mathbb{R}} \leq D\|\nabla q\|_{W_{m+k}^m(\Omega)} . \tag{3.7a}$$

If $k \geq 0$, there exists a unique real constant c and a unique s in $W_{m+k}^{m+1}(\Omega)$ such that q has the decomposition

$$q = c + s \, ,$$

and

$$\|s\|_{W_{m+k}^{m+1}(\Omega)} \leq D\|\nabla q\|_{W_{m+k}^m(\Omega)} . \tag{3.7b}$$

When $m \leq -1$, the constant D is equal to $\mathcal{B}$, the constant of (3.4) with parameters $-m$ and $-k$.

Proof : Consider first the case where $m \geq 0$. Let q be a distribution of $\mathcal{D}'(\Omega)$ such that ∇q belongs to $W_{m+k}^m(\Omega)^3$. We propose to extend q to $\mathbb{R}^3$ because the statement of the theorem is true in $\mathbb{R}^3$. With the notation introduced in the proof of Lemma 3.1, q is a distribution of $\mathcal{D}'(\Omega_{R_0})$ and ∇q belongs to $H^m(\Omega_{R_0})^3$. Since Ω_{R_0} is a bounded domain, the last property implies that q belongs to $H^{m+1}(\Omega_{R_0})$ (*cf.* for instance the references in [1]). By Calderón [6], q has an H^{m+1} extension $\tilde{q}$ in Ω' such that the function

$$\bar{q} = \begin{cases} \tilde{q} & \text{in } \Omega', \\ q & \text{in } \Omega \end{cases}$$

belongs to $H^{m+1}(B_{R_0})$ and $\nabla \bar{q}$ belongs to $W_{m+k}^m(\mathbb{R}^3)^3$. Then, applying Proposition 2.2 when $k \leq -1$, we find that $\bar{q}$ belongs to $W_{m+k}^{m+1}(\mathbb{R}^3)$ and applying Corollary 2.9 when $k \geq 0$, we find that $\bar{q} = s + c$ with $c \in \mathbb{R}$ and $s \in W_{m+k}^{m+1}(\mathbb{R}^3)$. In the last case, the uniqueness of s and c follows readily from the fact that the constants do not belong to $W_{m+k}^{m+1}(\Omega)$. In either case, the equivalence of norms (1.5) yields (3.7a) or (3.7b).

The above extension is not obvious when $m \leq -1$, but we can instead use an indirect argument. More precisely we can construct an element of $W_{m+k}^{m+1}(\Omega)$ with the same gradient as q. Set $\vec{\ell} = \nabla q$. Then, $\vec{\ell}$ vanishes on $\mathcal{V}(\Omega)$ and by virtue of the density Theorem 3.4, $\vec{\ell}$ vanishes on $V_{-(m+k)}^{-m}(\Omega)$. Therefore, $\vec{\ell}$ belongs to $\left[V_{-(m+k)}^{-m}(\Omega)\right]^0$ and the second isomorphism of Theorem 3.2 implies that there exists s in $W_{m+k}^{m+1}(\Omega)$ such that

$$\vec{\ell} = \nabla s \, ,$$

with

$$\|s\|_{W_{m+k}^{m+1}(\Omega)} \leq \mathcal{B}\|\vec{\ell}\|_{W_{m+k}^m(\Omega)} \quad \text{if } k \geq 0,$$

or
$$\|s\|_{W^{m+1}_{m+k}(\Omega)/\mathbb{R}} \leq B\|\vec{\ell}\|_{W^m_{m+k}(\Omega)} \quad \text{if } k \leq -1.$$

In both cases, q and s differ by a constant since they have the same gradient. When $k \leq -1$, the constant functions belong to $W^{m+1}_{m+k}(\Omega)$ and therefore, q also belongs to this space and satisfies (3.7a). When $k \geq 0$, the constant functions do not belong to $W^{m+1}_{m+k}(\Omega)$ and q is necessarily of the form $s + c$. $\diamond$

We end this section by lifting the divergence of functions that vanish on the boundary; this completes Lemma 3.1. We introduce the space, for $m \geq 1$:

$$X_0 = \left\{ \vec{v} \in [W^m_{m+k}(\Omega) \cap \overset{\circ}{W}{}^1_{1+k}(\Omega)]^3 \,;\, \text{div } \vec{v} = 0 \right\}.$$

Theorem 3.6. *Let m and k be two integers with k in $\mathbb{Z}$ and $m \geq 2$. Assume that Ω' is of class $C^{m-1,1}$. The divergence operator is an isomorphism from the quotient space $[W^m_{m+k}(\Omega) \cap \overset{\circ}{W}{}^1_{1+k}(\Omega)]^3 / X_0$ onto $W^{m-1}_{m+k}(\Omega)$ if $k \leq 0$ or onto $W^{m-1}_{m+k}(\Omega) \perp \mathbb{R}$ if $k \geq 1$.*

Remark 3.7. We start with $m = 2$ because the case $m = 1$ is proved by Theorem 3.2. $\diamond$

Proof : Let z belong to $W^{m-1}_{m+k}(\Omega)$ if $k \leq 0$ or $W^{m-1}_{m+k}(\Omega) \perp \mathbb{R}$ when $k \geq 1$.
1. Let us extend z in Ω' in such a way that the extended function has mean value zero in each connected component Ω'_i. First, as z belongs to H^{m-1} in a neighbourhood of Γ, it has an H^{m-1} extension in Ω' and the extended function (that we still denote by z) belongs to $W^{m-1}_{m+k}(\mathbb{R}^3)$. Next, in each Ω'_i, we fix a basis function θ_i in $\mathcal{D}(\Omega'_i)$ that satisfies

$$\int_{\Omega'_i} \theta_i \, d\mathbf{x} = 1.$$

Let $c_i = \int_{\Omega'_i} z \, d\mathbf{x}$ and consider the extension of z:

$$\bar{z} = z \quad \text{in } \Omega \quad \text{and} \quad \bar{z} = z - \sum_{i=1}^{I} c_i \theta_i \quad \text{in } \Omega'.$$

Clearly, $\bar{z}$ belongs to $W^{m-1}_{m+k}(\mathbb{R}^3)$,

$$\int_{\Omega'_i} \bar{z} \, d\mathbf{x} = \int_{\Omega'_i} (z - c_i \theta_i) \, d\mathbf{x} = 0$$

and

$$\|\bar{z}\|_{W^{m-1}_{m+k}(\mathbb{R}^3)} \leq C_1 \|z\|_{W^{m-1}_{m+k}(\Omega)}.$$

In addition, when $k \geq 1$, $\int_{\mathbb{R}^3} \bar{z} \, d\mathbf{x} = 0$, *i.e.* $\bar{z}$ is orthogonal to constants.

2. Now, it follows from Theorem 2.3 that we can find a function $\vec{\varphi}$ in $W_{m+k}^m(\mathbb{R}^3)^3$ such that

$$\operatorname{div}\vec{\varphi} = \bar{z} \quad \text{and} \quad \|\vec{\varphi}\|_{W_{m+k}^m(\mathbb{R}^3)} \le C_2 \|\bar{z}\|_{W_{m+k}^{m-1}(\mathbb{R}^3)}\,.$$

In particular, $\vec{\varphi}$ satisfies

$$\forall 1 \le i \le I\,, \quad \int_{\Gamma_i} \vec{\varphi}\cdot\vec{n}\,d\sigma = \int_{\Omega_i'} \operatorname{div}\vec{\varphi}\,d\mathbf{x} = \int_{\Omega_i'} \bar{z}\,d\mathbf{x} = 0. \tag{3.8}$$

It remains to construct an adequate divergence-free function that takes the same value as $\vec{\varphi}$ on Γ. More precisely, with the notations of Lemma 3.1, we look for a divergence-free function $\vec{w}$ in $H^m(\Omega_{R_0})^3$ such that $\gamma_0\vec{w} = \gamma_0\vec{\varphi}$ on Γ and $\gamma_0\vec{w} = \gamma_1\vec{w} = \cdots = \gamma_{m-1}\vec{w} = \vec{0}$ on ∂B_{R_0}, this last condition allowing to extend $\vec{w}$ by zero outside B_{R_0} without loss of regularity. Note that only the first trace of $\vec{w}$ is imposed on Γ and that, owing to (3.8), it satisfies the first compatibility condition of Héron [22]. Then it is easy to choose the higher order traces $\gamma_1\vec{w},\dots,\gamma_{m-1}\vec{w}$ on Γ so that the remaining compatibility conditions of Héron for the existence of a divergence-free lift are satisfied (*cf.* also Amrouche & Girault [2]) and

$$\|\vec{w}\|_{H^m(\Omega_{R_0})} \le C_3 \|\vec{\varphi}\|_{W_{m+k}^m(\mathbb{R}^3)}\,.$$

Then the function $\vec{v} = \vec{\varphi} - \vec{w}$ belongs to $[W_{m+k}^m(\Omega)\cap \overset{\circ}{W}{}^1_{1+k}(\Omega)]^3$ and satisfies

$$\operatorname{div}\vec{v} = z \quad \text{in } \Omega\,,$$

$$\|\vec{v}\|_{W_{m+k}^m(\Omega)} \le C_4 \|z\|_{W_{m+k}^{m-1}(\Omega)}\,. \qquad \diamondsuit$$

Remark 3.8. In contrast with Lemma 3.1, Theorem 3.6 does not give rise to a useful isomorphism for the gradient operator, because the dual of $W_{m+k}^m(\Omega)\cap \overset{\circ}{W}{}^1_{1+k}(\Omega)$ is not a convenient space. $\qquad\qquad \diamondsuit$

4. Vector potentials

In this section, we answer the following questions: given a divergence-free vector field in $W_{m+k}^m(\Omega)^3$, does it always have a divergence-free vector potential? If it does, is the vector potential unique? If it is not unique, can uniqueness be achieved by imposing adequate boundary conditions?

The first theorem answers the question of existence.

Theorem 4.1. *Let m belong to $\mathbb{N}$ and k belong to $\mathbb{Z}$. Assume that Ω' is Lipschitz-continuous when $m = 0$ and of class $C^{m-1,1}$ when $m \geq 1$. For any function $\vec{v}$ in $W^m_{m+k}(\Omega)^3$ if $m \geq 1$ or in $H_k(\mathrm{div}; \Omega)$ if $m = 0$ that satisfies:*

$$\mathrm{div}\, \vec{v} = 0 \quad \text{in } \Omega \quad \text{and} \quad \langle \vec{v} \cdot \vec{n}, 1 \rangle_{\Gamma_i} = 0 \quad \text{for } 1 \leq i \leq I, \quad (4.1)$$

there exists $\vec{\varphi}$ in $W^{m+1}_{m+k}(\Omega)^3$ such that

$$\vec{v} = \mathbf{curl}\, \vec{\varphi} \quad \text{and} \quad \mathrm{div}\, \vec{\varphi} = 0 \quad \text{in } \Omega. \quad (4.2)$$

Conversely, if Ω' is Lipschitz-continuous, any function $\vec{v} = \mathbf{curl}\, \vec{\varphi}$ with $\vec{\varphi}$ in $W^{m+1}_{m+k}(\Omega)^3$ satisfies (4.1).

Remark 4.2. The notation $\langle \cdot, \cdot \rangle_{\Gamma_i}$ stands for the duality between $H^{-\frac{1}{2}}(\Gamma_i)$ and $H^{\frac{1}{2}}(\Gamma_i)$. The space $H^{-\frac{1}{2}}(\Gamma_i)$ is well-defined because Γ_i is a closed surface. $\quad \diamond$

Proof : The proof proceeds in three steps.

1. Let us first prove the converse. Obviously, $\vec{v} = \mathbf{curl}\, \vec{\varphi}$ is divergence-free. Since $m + 1 \geq 1$, $\vec{\varphi}$ is H^1 in a neighbourhood of each Γ_i; therefore, $\vec{\varphi}$ has an H^1 extension, still denoted by $\vec{\varphi}$, in each connected component Ω'_i. Applying Green's formula in Ω'_i, we obtain

$$0 = \int_{\Omega'_i} \mathrm{div}\, (\mathbf{curl}\, \vec{\varphi})\, d\mathbf{x} = \langle \mathbf{curl}\, \vec{\varphi} \cdot \vec{n}, 1 \rangle_{\Gamma_i} = \langle \vec{v} \cdot \vec{n}, 1 \rangle_{\Gamma_i}, \quad \text{for } 1 \leq i \leq I,$$

where $\vec{n}$ denotes the normal vector exterior to Ω'_i and therefore interior to Ω. Note that, even when $I = 1$, a divergence-free $\vec{v}$ does not necessarily satisfy this boundary condition if $k \leq 0$.

2. Now, to construct the vector potential, we first extend $\vec{v}$ in Ω' in such a way that it keeps the same regularity and stays divergence-free. The construction depends on the value of m.

a. If $m = 0$, we solve the following Neumann problem in each Ω'_i:

$$\Delta \theta = 0 \quad \text{in } \Omega'_i \quad \text{and} \quad \frac{\partial \theta}{\partial n} = \vec{v} \cdot \vec{n} \quad \text{on } \Gamma_i.$$

Owing to the boundary condition in (4.1), this problem has a unique solution in $H^1(\Omega'_i)/\mathbb{R}$ and

$$|\theta|_{H^1(\Omega'_i)} \leq C_1 \|\vec{v} \cdot \vec{n}\|_{H^{-\frac{1}{2}}(\Gamma_i)}.$$

Let us take

$$\vec{w} = \nabla \theta \quad \text{in } \Omega' \quad \text{and} \quad \vec{w} = \vec{v} \quad \text{in } \Omega.$$

Then $\vec{w}$ belongs to $H_k(\mathrm{div}; \mathbb{R}^3)$, $\mathrm{div}\, \vec{w} = 0$ in $\mathbb{R}^3$ and

$$\|\vec{w}\|_{H_k(\mathrm{div};\mathbb{R}^3)} \leq C_2 \|\vec{v}\|_{H_k(\mathrm{div};\Omega)}.$$

b. If $m \geq 1$, $\vec{v}$ has traces up to the order $m-1$ on each Γ_i. Therefore, the boundary conditions (4.1) imply that the traces $\gamma_0 \vec{v}, \gamma_1 \vec{v}, \ldots, \gamma_{m-1} \vec{v}$ satisfy the compatibility conditions of Héron [22] that allow to lift those traces in each Ω_i' by a divergence-free function in $H^m(\Omega_i')^3$. This function is of course not unique, but a representative can be chosen so that it depends continuously on $\vec{v}$. Therefore, denoting again the extended function by $\vec{w}$, we have the same conclusion as above: $\vec{w}$ belongs to $W_{m+k}^m(\mathbb{R}^3)^3$, $\operatorname{div} \vec{w} = 0$ in $\mathbb{R}^3$ and

$$\|\vec{w}\|_{W_{m+k}^m(\mathbb{R}^3)} \leq C_3 \|\vec{v}\|_{W_{m+k}^m(\Omega)} \,.$$

3. Finally, to obtain the vector potential in $\mathbb{R}^3$, the discussion splits according to the value of k.

a. If $k \leq 2$, Theorem 2.5 shows that $\vec{w}$ has a vector potential $\vec{\varphi}$ in $W_{m+k}^{m+1}(\mathbb{R}^3)^3$ that satisfies

$$\vec{w} = \operatorname{\mathbf{curl}} \vec{\varphi} \quad \text{and} \quad \operatorname{div} \vec{\varphi} = 0 \quad \text{in } \mathbb{R}^3 \,,$$

and by choosing an adequate representative, it depends continuously on $\vec{w}$:

$$\|\vec{\varphi}\|_{W_{m+k}^{m+1}(\mathbb{R}^3)} \leq C_4 \|\vec{w}\|_{W_{m+k}^m(\mathbb{R}^3)} \,.$$

b. When $k \geq 3$, the proof of Theorem 2.5 fails because it relies on the solution of (2.11) (with right-hand side $\operatorname{\mathbf{curl}} \vec{w}$) and this solution exists in $W_{m+k}^{m+1}(\mathbb{R}^3)^3$ only if $\operatorname{\mathbf{curl}} \vec{w}$ is orthogonal to all polynomials of $(\mathcal{P}_{k-1}^{\Delta})^3$. Usually, $\vec{w}$ does not satisfy this orthogonality condition; but $\vec{w}$ is only one possible extension of $\vec{v}$ and we are going to show now how to construct another extension that satisfies it.

Let us fix one connected component of Ω', say Ω_1', and let D_1 and D_2 be two concentric balls, with radii $0 < r_1 < r_2$, such that $D_1 \subset D_2 \Subset \Omega_1'$. Let a be a cut-off function, similar to that introduced in the proof of Theorem 3.4:

$$a \in \mathcal{D}(\mathbb{R}^3) \quad, \quad 0 \leq a(\mathbf{x}) \leq 1 \quad \forall \mathbf{x} \in \mathbb{R}^3 \quad, \quad a \equiv 1 \text{ in } D_1 \quad \text{and} \quad \operatorname{supp}(a) \subset D_2 \,.$$

Then the mapping

$$q \mapsto \left\{ \int_{\Omega_1'} a(\mathbf{x}) q(\mathbf{x})^2 \, d\mathbf{x} \right\}^{\frac{1}{2}}$$

is a norm on the polynomial space $\mathbb{P}_n$ for any n. Now, let $\vec{w}$ be the above extension of $\vec{v}$ and let $\vec{\varphi}$ be one of its vector potentials constructed in Lemma 3.3. Let $\vec{p}$ in $\mathbb{P}_{k-3}^3$ be the solution of the system

$$\forall \vec{q} \in \mathbb{P}_{k-3}^3 \quad, \quad \int_{\Omega_1'} a \vec{p} \cdot \vec{q} \, d\mathbf{x} = \int_{\mathbb{R}^3} \vec{\varphi} \cdot \vec{q} \, d\mathbf{x} \,.$$

The right-hand side is bounded because $\rho(r)^{k-3}\vec{\varphi}$ is in $L^1(\mathbb{R}^3)^3$ and the system has exactly one solution by virtue of the above norm. Then $\vec{p}$ satisfies in particular

$$\forall \vec{q} \in (\mathcal{P}^\Delta_{k-1})^3 \quad , \quad \int_{\Omega'_1} a\,\vec{p} \cdot \mathbf{curl}\,\mathbf{curl}\,\vec{q}\,dx = \int_{\mathbb{R}^3} \vec{\varphi} \cdot \mathbf{curl}\,\mathbf{curl}\,\vec{q}\,dx .$$

Set $\vec{z} = \mathbf{curl}(a\vec{p})$; it is divergence-free, owing to the support of a, $\vec{z}$ belongs to $\mathcal{D}(\Omega'_1)^3$ and the above equation implies

$$\forall \vec{q} \in (\mathcal{P}^\Delta_{k-1})^3 \quad , \quad \int_{\Omega'_1} \mathbf{curl}\,\vec{z} \cdot \vec{q}\,dx = \int_{\mathbb{R}^3} \mathbf{curl}\,\vec{w} \cdot \vec{q}\,dx .$$

Then, we take the following extension of $\vec{v}$:

$$\vec{u} = \vec{w} - \vec{z} \quad \text{in } \Omega'_1 \quad \text{and} \quad \vec{u} = \vec{w} \quad \text{elsewhere} .$$

By construction, $\mathbf{curl}\,\vec{u}$ is orthogonal to all polynomials of $(\mathcal{P}^\Delta_{k-1})^3$ and

$$\|\vec{u}\|_{W^m_{m+k}(\mathbb{R}^3)} \leq C_5 \|\vec{w}\|_{W^m_{m+k}(\mathbb{R}^3)} ,$$

since a is a fixed function and $\vec{p}$ belongs to a finite-dimensional space. Then an easy argument (similar to that used in proving Theorem 2.5) shows that $\vec{u}$ has a unique divergence-free vector potential $\vec{\lambda}$ in $W^{m+1}_{m+k}(\mathbb{R}^3)^3$ and

$$\|\vec{\lambda}\|_{W^{m+1}_{m+k}(\mathbb{R}^3)} \leq C_6 \|\vec{u}\|_{W^m_{m+k}(\mathbb{R}^3)} . \qquad \diamond$$

Clearly the conditions (4.2) alone are not sufficient to ensure uniqueness of the vector potential; other conditions must be imposed. In the case of a simply-connected bounded domain, one usually forces either the normal component to vanish or the tangential components to vanish and this is achieved by solving respectively a Neumann problem or a Dirichlet problem for the Laplace operator in this domain. To apply the same approach here, we must know how to solve these two problems in an exterior domain. For the sake of briefness, we shall only work with the tangential components here; the results obtained can be easily adapted to the normal components. The results that we state below, concerning the Dirichlet problem, are established by Giroire in [17].

Lemma 4.3. *Let m belong to $\mathbb{N}$ and suppose that the domain Ω' is of class $\mathcal{C}^{m,1}$. For any f in $W^{m-1}_m(\Omega)$ and any g in $H^{m+\frac{1}{2}}(\Gamma)$, there exists a unique function u in $W^{m+1}_m(\Omega)$ solution of*

$$-\Delta u = f \quad \text{in } \Omega \quad \text{and} \quad u = g \quad \text{on } \Gamma .$$

The mapping $(f, g) \mapsto u$ is an isomorphism from $W_m^{m-1}(\Omega) \times H^{m+\frac{1}{2}}(\Gamma)$ onto $W_m^{m+1}(\Omega)$.

In contrast with a bounded domain, the Dirichlet problem for the Laplace operator with zero data can have nontrivial solutions in an exterior domain; it depends on the exponent of the weight. This possibility is described by the following definition.

Definition 4.4. For any integer $k \geq 1$, we denote by $\mathcal{A}_{k-1}^{\Delta}$ the subspace of all functions $v(p)$ in $W_{-k}^1(\Omega)$ of the form $v(p) = w(p) - p$, where p runs over all polynomials of $\mathcal{P}_{k-1}^{\Delta}$ and $w(p)$ is the unique solution in $W_0^1(\Omega)$ of the Dirichlet problem

$$\Delta w(p) = 0 \quad \text{in } \Omega \quad \text{and } w(p) = p \quad \text{on } \Gamma. \qquad \Diamond$$

By convention, we set $\mathcal{A}_{k-1}^{\Delta} = \{0\}$ when $k \leq 0$. Note that $\mathcal{A}_{k-1}^{\Delta}$ is a finite-dimensional space of the same dimension as $\mathcal{P}_{k-1}^{\Delta}$. Also, owing to the regularity result of Lemma 4.3, we see that when Ω' is of class $\mathcal{C}^{m,1}$ for some positive integer m, then the functions of $\mathcal{A}_{k-1}^{\Delta}$ belong to $W_{m-k}^{m+1}(\Omega)$. In fact, $\mathcal{A}_{k-1}^{\Delta}$ is the null space of the Dirichlet problem for the Laplace operator in $W_{-k}^1(\Omega)$, which means that in this space, the solution is not unique. This permits to extend the statement of Lemma 4.3.

Theorem 4.5. Let m and k belong to $\mathbb{N}$ and let Ω' be of class $\mathcal{C}^{m,1}$. For any f in $W_{m-k}^{m-1}(\Omega)$ and any g in $H^{m+\frac{1}{2}}(\Gamma)$, there exists a unique function u in $W_{m-k}^{m+1}(\Omega)/\mathcal{A}_{k-1}^{\Delta}$ solution of

$$-\Delta u = f \quad \text{in } \Omega \quad \text{and} \quad u = g \quad \text{on } \Gamma.$$

The mapping $(f, g) \mapsto u$ is an isomorphism from $W_{m-k}^{m-1}(\Omega) \times H^{m+\frac{1}{2}}(\Gamma)$ onto the quotient space $W_{m-k}^{m+1}(\Omega)/\mathcal{A}_{k-1}^{\Delta}$.

By duality, Theorem 4.5 implies that the Dirichlet problem for the Laplace operator has a solution in $W_k^1(\Omega)$ for $k \geq 1$ if and only if its data satisfy some orthogonality relation. It is specified in the next result.

Theorem 4.6. Let m and k belong to $\mathbb{N}$ with $k \geq 1$ and let Ω' be of class $\mathcal{C}^{m,1}$. For f given in $W_{m+k}^{m-1}(\Omega)$ and g in $H^{m+\frac{1}{2}}(\Gamma)$, the problem

$$-\Delta u = f \quad \text{in } \Omega \quad \text{and} \quad u = g \quad \text{on } \Gamma$$

has a (unique) solution u in $W_{m+k}^{m+1}(\Omega)$ if and only if

$$\forall v(p) \in \mathcal{A}_{k-1}^{\Delta}, \quad \int_{\Omega} f \, v(p) \, d\mathbf{x} = \langle g, \frac{\partial v(p)}{\partial n} \rangle_{\Gamma}. \qquad (4.3)$$

The next proposition shows how to fix the vector potential in a simplified case, by imposing a condition on its tangential trace. But beforehand, we require the following auxiliary lemma.

Lemma 4.7. *Assume that Ω' is Lipschitz-continuous and let $v(1)$ be the function of Definition 4.4. Then $v(1)$ spans $\mathcal{A}_0^\Delta$ and*

$$\langle \frac{\partial v(1)}{\partial n}, 1 \rangle_\Gamma > 0.$$

Proof: Clearly, $v(1)$ spans $\mathcal{A}_0^\Delta$. By definition, $v(1) = w(1) - 1$, where $w(1) \in W_0^1(\Omega)$ is the unique solution of

$$\Delta w(1) = 0 \quad \text{in } \Omega \quad, \quad w(1) = 1 \quad \text{on } \Gamma.$$

An easy density argument shows that the following Green's formula holds

$$0 = \langle \Delta w(1), w(1) \rangle = -\int_\Omega |\nabla w(1)|^2 \, d\mathbf{x} + \langle \frac{\partial w(1)}{\partial n}, w(1) \rangle_\Gamma.$$

Hence

$$\langle \frac{\partial w(1)}{\partial n}, 1 \rangle_\Gamma = \int_\Omega |\nabla w(1)|^2 \, d\mathbf{x} > 0,$$

as $w(1)$ cannot be a constant. $\diamond$

Proposition 4.8. *Let $m \geq 0$ and $k \leq 2$ be two integers and assume that Ω' is of class $C^{m+1,1}$, simply-connected and with a connected boundary (i.e. $I = 1$). Each function $\vec{v}$ in $V_{m+k}^m(\Omega)$ has a unique vector potential $\vec{\psi}$ in $W_{m+k}^{m+1}(\Omega)^3$ such that*

$$\vec{v} = \mathbf{curl}\, \vec{\psi} \quad, \quad div\, \vec{\psi} = 0 \quad \text{in } \Omega \quad, \quad \vec{\psi} \times \vec{n} = \vec{0} \quad \text{on } \Gamma, \tag{4.4}$$

$$\int_\Gamma \vec{\psi} \cdot \vec{n} \, d\sigma = 0, \tag{4.5}$$

and if $k \leq -1$,

$$\forall q \in \mathcal{P}_{-k}^\Delta, \quad \int_\Gamma (\vec{\psi} \cdot \vec{n}) q \, d\sigma = 0. \tag{4.6}$$

It depends continuously on $\vec{v}$:

$$\|\vec{\psi}\|_{W_{m+k}^{m+1}(\Omega)} \leq C \|\vec{v}\|_{W_{m+k}^m(\Omega)}. \tag{4.7}$$

When $k \geq 3$, there does not always exist a vector potential $\vec{\psi}$ in $W_{m+k}^{m+1}(\Omega)^3$ that satisfies (4.4).

Proof : Take $k \leq 2$. First, we extend $\vec{v}$ by zero in Ω'. The extended function $\vec{v}$ belongs to $V_{m+k}^m(\mathbb{R}^3)$; its corresponding vector potential $\vec{\varphi}$ belongs to $V_{m+k}^{m+1}(\mathbb{R}^3)$ and satisfies in particular $\mathbf{curl}\,\vec{\varphi} = \vec{0}$ in Ω'. Since Ω' is simply-connected, this implies that,

$$\vec{\varphi} = \nabla \mu \quad \text{in } \Omega' \quad \text{for some function } \mu \in H^{m+2}(\Omega')/\mathbb{R}\,.$$

Let μ_0 be the representative of μ with zero mean value in Ω' and let us solve the Dirichlet problem in $W_0^1(\Omega)$:

$$\Delta z_0 = 0 \quad \text{in } \Omega \quad \text{and} \quad z_0 = \mu_0 \quad \text{on } \Gamma. \tag{4.8}$$

Now, the discussion splits according to the value of k. When $k = 2$, we take

$$\vec{\psi} = \vec{\varphi} - \nabla z, \quad \text{where } z = z_0 + c_1 w(1),$$

and the constant c_1 is chosen to satisfy

$$\langle z, \frac{\partial v(1)}{\partial n} \rangle_\Gamma = 0. \tag{4.9}$$

When $k = 1$, we take $\vec{\psi}$ of the same form, but we define c_1 so that $\vec{\psi}$ satisfies (4.5). Note that in either case Lemma 4.7 shows that (4.9) or (4.5) determines c_1 uniquely. Moreover, when $k = 2$, (4.9) is the compatibility condition (4.3) of Theorem 4.6 that guarantees that z belongs to $W_1^1(\Omega)$. In this case, (4.5) is automatically satisfied. Therefore, owing that Ω' is of class $C^{m+1,1}$, it follows from Theorem 4.6 that z belongs to $W_{m+k}^{m+2}(\Omega)$ for $k = 1$ or 2.

When $k \le -1$, the conditions (4.4) and (4.5) are not sufficient to guarantee uniqueness, as we shall see below. In this case, we take $\vec{\psi}$ of the form

$$\vec{\psi} = \vec{\varphi} - \nabla z_0 - \nabla(w(p) - p), \tag{4.10}$$

where the polynomial p in $\mathcal{P}_{-k}^\Delta$, is chosen to satisfy the conditions (4.6). This amounts to solve a linear system of dimension $N(-k)$, the dimension of $\mathcal{P}_{-k}^\Delta$. To prove that this system has a unique solution, it suffices to prove that if there exists a polynomial p in $\mathcal{P}_{-k}^\Delta$ such that

$$\forall q \in \mathcal{P}_{-k}^\Delta, \quad \int_\Gamma \frac{\partial}{\partial n}(w(p) - p)\, q\, d\sigma = 0\,,$$

then $p = 0$. This system implies in particular

$$\int_\Gamma \frac{\partial w(p)}{\partial n}\, w(p)\, d\sigma = \int_\Gamma \frac{\partial p}{\partial n}\, p\, d\sigma\,.$$

Then the argument used in proving Lemma 4.7 gives

$$\|\nabla w(p)\|_{L^2(\Omega)}^2 = \int_\Gamma \frac{\partial w(p)}{\partial n}\, w(p)\, d\sigma = \int_\Gamma \frac{\partial p}{\partial n}\, p\, d\sigma\,.$$

But since p is harmonic, Green's formula yields

$$\|\nabla p\|^2_{L^2(\Omega')} = -\int_\Gamma \frac{\partial p}{\partial n}\, p\, d\sigma\,,$$

where the minus sign takes into account the fact that the normal vector $\vec{n}$ points outside Ω and therefore inside Ω'. Putting together these two equations, we obtain

$$\|\nabla w(p)\|^2_{L^2(\Omega)} + \|\nabla p\|^2_{L^2(\Omega')} = 0\,.$$

Since $w(p)$ belongs to $W^1_0(\Omega)$, this means that $w(p) = 0$ and therefore $p = 0$.

The uniqueness of $\vec{\psi}$ follows easily from this construction. We prove uniqueness for the general case $k \leq 0$, the other cases being simpler. Let $\vec{v}$ have two vector potentials that satisfy (4.4), (4.5) and (4.6). As Ω is simply-connected, their difference, say $\vec{\theta}$, satisfies $\vec{\theta} \in W^{m+1}_{m+k}(\Omega)^3$ and $\vec{\theta} = \nabla h$ with

$$\Delta h = 0 \quad \text{in } \Omega\,, \quad h = c_1 \text{ on } \Gamma \text{ and } \forall q \in \mathcal{P}^\Delta_{-k} \quad \int_\Gamma \frac{\partial h}{\partial n}\, q\, d\sigma = 0\,, \qquad (4.11)$$

where c_1 is a constant.

First, Theorem 3.5 shows that h belongs to $W^{m+2}_{m+k}(\Omega)$. Therefore h is necessarily of the form $h = c_1 + v(p)$, where $v(p)$ is some element of the kernel $\mathcal{A}^\Delta_{-k}$. Then the conditions (4.11) show that $p = 0$ and hence $\vec{\theta} = \vec{0}$.

In each case, the continuity inequality (4.7) stems readily from the above construction.

Finally, it is easy to see that, as soon as $k \geq 3$, one constant c_1 is not sufficient to enforce the necessary compatibility conditions of Theorem 4.6 so that it is not possible to find $\vec{\psi}$ in $W^{m+1}_{m+k}(\Omega)^3$ that satisfies (4.4). $\diamond$

Remark 4.9. The statement of proposition 4.8 can be extended to the case where the boundary Γ is not connected, but the proof (which is written in [13]) is a little more technical. $\diamond$

When $m \geq 1$, the construction of Proposition 4.8 does not apply directly to functions that only satisfy $\vec{v} \cdot \vec{n} = 0$ on the boundary. The reason is that if such functions are extended by zero outside Ω, they are no longer in $W^m_{m+k}(\mathbb{R}^3)^3$. But in fact, this construction is not necessary; what we really require is a regularity result. This is the object of the next theorem.

Theorem 4.10. *Let $m \geq 1$ and $k \leq 2$ be two integers and assume that Ω' is of class $C^{m+1,1}$, simply-connected and with a connected boundary. The conclusion of Proposition 4.8 is valid for all functions $\vec{v}$ in $W^m_{m+k}(\Omega)^3$ that satisfy $\text{div}\,\vec{v} = 0$ in Ω and $\vec{v} \cdot \vec{n} = 0$ on Γ.*

Proof As $m \geq 1$, we can apply the conclusion of Proposition 4.8 with $m = 0$. Thus, $\vec{v}$ has a unique vector potential $\vec{\psi}$ in $W^1_k(\Omega)^3$ that satisfies (4.4), (4.5) and (4.6) and it remains to establish that $\vec{\psi}$ belongs in fact to $W_{m+k}^{m+1}(\Omega)^3$. This regularity result can be proved by induction, using an adequate partition of unity.

Assume that $\vec{\psi}$ belongs to $W_{\ell+k}^{\ell+1}(\Omega)^3$ for some ℓ with $0 \leq \ell \leq m-1$ and let us prove that $\vec{\psi}$ belongs to $W_{\ell+1+k}^{\ell+2}(\Omega)^3$. With the notations of Lemma 3.1, let λ and μ be two scalar, nonnegative functions in $\mathcal{C}^\infty(\mathbb{R}^3)$ that satisfy

$$\forall \mathbf{x} \in B_{R_0}, \quad \lambda(\mathbf{x}) = 1, \quad \text{supp}(\lambda) \subset B_{R_0+1} \quad , \quad \forall \mathbf{x} \in \mathbb{R}^3, \quad \lambda(\mathbf{x}) + \mu(\mathbf{x}) = 1.$$

Let Ω_{R_0+1} denote the intersection $\Omega \cap B_{R_0+1}$ and let C_{R_0} denote the exterior (i.e. the complement) of B_{R_0}. Then, we can write

$$\vec{v} = \mathbf{curl}(\lambda\vec{\psi}) + \mathbf{curl}(\mu\vec{\psi})$$

and in view of their supports, it suffices to examine $\lambda\vec{\psi}$ on Ω_{R_0+1} and $\mu\vec{\psi}$ on C_{R_0}. Consider first the function $\lambda\vec{\psi}$. Owing to the induction hypothesis, it satisfies

$$\mathbf{curl}(\lambda\vec{\psi}) = \lambda\vec{v} - \nabla\mu \times \vec{\psi} \in H^{\ell+1}(\Omega_{R_0+1})^3 \,,$$

$$\text{div}(\lambda\vec{\psi}) = \nabla\lambda \cdot \vec{\psi} \in H_0^{\ell+1}(\Omega_{R_0+1}) \quad \text{and} \quad \int_{\Omega_{R_0+1}} \text{div}(\lambda\vec{\psi}) \, d\mathbf{x} = 0 \,.$$

It follows from [1] and this last property that there exists $\vec{w}$ in $H_0^{\ell+2}(\Omega_{R_0+1})^3$ such that

$$\text{div}\,\vec{w} = \nabla\lambda \cdot \vec{\psi} \quad \text{and} \quad \|\vec{w}\|_{H^{\ell+2}(\Omega_{R_0+1})} \leq C_1 \|\vec{\psi}\|_{H^{\ell+1}(\Omega_{R_0+1})} \leq C_2 \|\vec{v}\|_{W_{m+k}^m(\Omega)} \,.$$

Then the function $\vec{z} = \lambda\vec{\psi} - \vec{w}$ satisfies

$$\mathbf{curl}\,\vec{z} \in H^{\ell+1}(\Omega_{R_0+1})^3 \quad , \quad \text{div}\,\vec{z} = 0 \ \text{in} \ \Omega_{R_0+1} \quad , \quad \vec{z} \times \vec{n} = \vec{0} \ \text{on} \ \partial\Omega_{R_0+1} \,,$$

$$\int_{\partial B_{R_0+1}} \vec{z} \cdot \vec{n} \, d\sigma = 0 \quad \text{and} \quad \int_\Gamma \vec{z} \cdot \vec{n} \, d\sigma = 0 \,.$$

Since the set Ω_{R_0+1} is bounded, simply-connected and has the same regularity as Ω, these properties imply [8] that $\vec{z}$ belongs to $H^{\ell+2}(\Omega_{R_0+1})^3$ with

$$\|\vec{z}\|_{H^{\ell+2}(\Omega_{R_0+1})} \leq C_3 \|\mathbf{curl}\,\vec{z}\|_{H^{\ell+1}(\Omega_{R_0+1})} \leq C_4 \|\vec{v}\|_{W_{m+k}^m(\Omega)} \,.$$

Hence $\lambda\vec{\psi}$ also belongs to $H^{\ell+2}(\Omega_{R_0+1})^3$ and

$$\|\lambda\vec{\psi}\|_{H^{\ell+2}(\Omega_{R_0+1})} \leq C_5 \|\vec{v}\|_{W_{m+k}^m(\Omega)} \,.$$

Now, we consider $\mu\vec{\psi}$. It satisfies

$$\mathbf{curl}(\mu\vec{\psi}) = \mu\vec{v} - \nabla\lambda \times \vec{\psi} \in V_{\ell+1+k}^{\ell+1}(C_{R_0}) \quad , \quad \mathrm{div}(\mu\vec{\psi}) = \nabla\mu \cdot \vec{\psi} \in \overset{\circ}{W}_{\ell+1+k}^{\ell+1}(C_{R_0}),$$

owing to the induction hypothesis and the fact that $\nabla\lambda$ and $\nabla\mu$ have bounded supports. In addition, when $k = 2$, $\mathrm{div}(\mu\vec{\psi})$ is orthogonal to constants in C_{R_0} because $\mu\vec{\psi}$ vanishes on ∂C_{R_0}. Therefore, it follows from Theorem 3.2 that there exists $\vec{w}$ in $\overset{\circ}{W}_{\ell+1+k}^{\ell+2}(C_{R_0})^3$ such that

$$\mathrm{div}\,\vec{w} = \nabla\mu \cdot \vec{\psi},$$

$$\|\vec{w}\|_{W_{\ell+1+k}^{\ell+2}(C_{R_0})} \leq C_6\|\nabla\mu \cdot \vec{\psi}\|_{W_{\ell+1+k}^{\ell+1}(C_{R_0})} \leq C_7\|\vec{\psi}\|_{W_{\ell+k}^{\ell+1}(\Omega)} \leq C_8\|\vec{v}\|_{W_{m+k}^{m}(\Omega)}.$$

Thus the function $\vec{z} = \mu\vec{\psi} - \vec{w}$ satisfies

$$\mathbf{curl}\,\vec{z} \in V_{\ell+1+k}^{\ell+1}(C_{R_0}) \quad , \quad \mathrm{div}\,\vec{z} = 0 \;\text{ in } C_{R_0} \quad , \quad \vec{z} = \vec{0} \;\text{ on } \partial C_{R_0}.$$

Then the uniqueness of the vector potential defined by Proposition 4.8 implies that $\vec{z}$ belongs to $W_{\ell+1+k}^{\ell+2}(C_{R_0})^3$ and

$$\|\vec{z}\|_{W_{\ell+1+k}^{\ell+2}(C_{R_0})} \leq C_9\|\mathbf{curl}\,\vec{z}\|_{W_{\ell+1+k}^{\ell+1}(C_{R_0})} \leq C_{10}\|\vec{v}\|_{W_{m+k}^{m}(\Omega)} \;;$$

consequently $\mu\vec{\psi}$ belongs also to $W_{\ell+1+k}^{\ell+2}(C_{R_0})^3$ and

$$\|\mu\vec{\psi}\|_{W_{\ell+1+k}^{\ell+2}(C_{R_0})} \leq C_{11}\|\vec{v}\|_{W_{m+k}^{m}(\Omega)}.$$

Hence $\vec{\psi}$ belongs to $W_{\ell+1+k}^{\ell+2}(\Omega)^3$ and the theorem follows by induction. $\qquad\qquad \diamond$

In addition to determining the vector potential, Theorem 4.10 permits to extend to exterior domains the important imbedding that we have used in the above proof in the bounded domain Ω_{R_0+1}. For any integers m in $I\!\!N$ and k in $Z\!\!\!Z$, we introduce the space

$$(X_N)_{m+k}^m(\Omega) = \{\vec{v} \in W_{m+k-1}^m(\Omega)^3 \,; \mathrm{div}\,\vec{v} \in W_{m+k}^m(\Omega), \mathbf{curl}\,\vec{v} \in W_{m+k}^m(\Omega)^3, \tag{4.12}$$
$$\vec{v} \times \vec{n} = \vec{0} \;\text{ on } \Gamma\}.$$

Corollary 4.11. *Let $m \geq 0$ and $k \leq 2$ be two integers and suppose that Ω' is of class $\mathcal{C}^{m+1,1}$. Then the space $(X_N)_{m+k}^m(\Omega)$ is continuously imbedded in $W_{m+k}^{m+1}(\Omega)^3$: there exists a constant C such that*

$$\forall\vec{\varphi} \in (X_N)_{m+k}^m(\Omega) ,$$

$$\|\vec{\varphi}\|_{W^{m+1}_{m+k}(\Omega)} \leq C\{\|\vec{\varphi}\|^2_{W^m_{m+k-1}(\Omega)} + \|div\,\vec{\varphi}\|^2_{W^m_{m+k}(\Omega)} + \|curl\,\vec{\varphi}\|^2_{W^m_{m+k}(\Omega)}\}^{\frac{1}{2}} \quad (4.13)$$

If in addition, Ω' is simply-connected and its boundary Γ is connected, there exists a constant C such that for all $\vec{\varphi}$ in $(X_N)^m_{m+k}(\Omega)$ we have:

$$\|\vec{\varphi}\|_{W^{m+1}_{m+k}(\Omega)} \leq C\{\|div\,\vec{\varphi}\|^2_{W^m_{m+k}(\Omega)} + \|\mathbf{curl}\,\vec{\varphi}\|^2_{W^m_{m+k}(\Omega)}$$

$$+ |\int_{\Gamma}(\vec{\varphi}\cdot\vec{n})\,d\sigma|^2 + \sum_{j=1}^{N(-k)} |\int_{\Gamma}(\vec{\varphi}\cdot\vec{n})q_j\,d\sigma|^2\}^{\frac{1}{2}}, \quad (4.14)$$

where the term $|\int_{\Gamma}(\vec{\varphi}\cdot\vec{n})\,d\sigma|^2$ can be dropped if $k \neq 1$ and where $\{q_j\}_{j=1}^{N(-k)}$ denotes a basis of $\mathcal{P}^{\Delta}_{-k}$. In other words, the seminorm in the right-hand side of (4.14) is a norm on $(X_N)^m_{m+k}(\Omega)$ equivalent to the norm $\|\vec{\varphi}\|_{W^{m+1}_{m+k}(\Omega)}$.

Proof : The proof is very similar to that of Theorem 4.10. With its notations, any $\vec{\varphi}$ in $(X_N)^m_{m+k}(\Omega)$ can be written $\vec{\varphi} = \lambda\vec{\varphi} + \mu\vec{\varphi}$, where $\lambda\vec{\varphi}$ has its support in Ω_{R_0+1} and $\mu\vec{\varphi}$ in C_{R_0}.

Consider first $\mu\vec{\varphi}$. Its domain C_{R_0} is simply-connected and has a very smooth and connected boundary ∂C_{R_0}. Next, $div(\mu\vec{\varphi})$ belongs to $W^m_{m+k}(C_{R_0})$ or $W^m_{m+k}(C_{R_0})\perp\mathbb{R}$ when $k = 2$, because $\mu\vec{\varphi}$ vanishes on ∂C_{R_0}. Therefore, applying Theorem 3.6, there exists $\vec{w}$ in $W^{m+1}_{m+k}(C_{R_0})^3$, that vanishes on ∂C_{R_0}, such that

$$div\,\vec{w} = div(\mu\vec{\varphi}),$$

$$\|\vec{w}\|_{W^{m+1}_{m+k}(C_{R_0})} \leq C_1\|div(\mu\vec{\varphi})\|_{W^m_{m+k}(C_{R_0})} \leq$$
$$\leq C_2\{\|div\,\vec{\varphi}\|_{W^m_{m+k}(C_{R_0})} + \|\vec{\varphi}\|_{H^m(B_{R_0+1}\setminus B_{R_0})}\}.$$

Let $\vec{\psi} = \mu\vec{\varphi} - \vec{w}$ and set $\vec{v} = \mathbf{curl}\,\vec{\psi}$. Then $\vec{v}$ belongs to $W^m_{m+k}(C_{R_0})^3$,

$$\|\vec{v}\|_{W^m_{m+k}(C_{R_0})} \leq C_3\{\|\vec{w}\|_{W^{m+1}_{m+k}(C_{R_0})} + \|\mathbf{curl}\,\vec{\varphi}\|_{W^m_{m+k}(C_{R_0})} + \|\vec{\varphi}\|_{H^m(B_{R_0+1}\setminus B_{R_0})}\},$$

and $\vec{\psi}$ satisfies (4.4), (4.5) and (4.6) in C_{R_0}. Therefore $\vec{\psi}$ is a vector potential of $\vec{v}$ in C_{R_0} and Theorem 4.10 and the uniqueness of the vector potential imply that $\vec{\psi}$ belongs to $W^{m+1}_{m+k}(C_{R_0})^3$ and

$$\|\vec{\psi}\|_{W^{m+1}_{m+k}(C_{R_0})} \leq C_4\|\vec{v}\|_{W^m_{m+k}(C_{R_0})}.$$

In turn, this implies that $\mu\vec{\varphi}$ belongs to $W^{m+1}_{m+k}(C_{R_0})^3$ and

$$\|\mu\vec{\varphi}\|_{W^{m+1}_{m+k}(C_{R_0})} \leq C_5\{\|div\,\vec{\varphi}\|_{W^m_{m+k}(C_{R_0})} +$$
$$+ \|\mathbf{curl}\,\vec{\varphi}\|_{W^m_{m+k}(C_{R_0})} + \|\vec{\varphi}\|_{H^m(B_{R_0+1}\setminus B_{R_0})}\}. \quad (4.15)$$

Finally, applying for example [8], the fact that $\lambda\vec{\varphi}$ has the bounded support Ω_{R_0+1}, of the same class as Ω', implies that $\lambda\vec{\varphi}$ belongs to $H^{m+1}(\Omega_{R_0+1})^3$ and

$$
\begin{aligned}
\|\lambda\vec{\varphi}\|_{H^{m+1}(\Omega_{R_0+1})} \leq C_6 \{ &\|\operatorname{div}(\lambda\vec{\varphi})\|_{H^m(\Omega_{R_0+1})} + \\
&+ \|\mathbf{curl}(\lambda\vec{\varphi})\|_{H^m(\Omega_{R_0+1})} + \|\lambda\vec{\varphi}\|_{H^m(\Omega_{R_0+1})} \} .
\end{aligned}
\tag{4.16}
$$

Combining these two results, we derive that $\vec{\varphi}$ belongs to $W^{m+1}_{m+k}(\Omega)^3$ and (4.13) follows easily from (4.15) and (4.16).

Finally, when Ω' is simply-connected and its boundary is connected, the inequality (4.14) is derived from (4.13) by reducing $\vec{\varphi}$ to a vector potential that satisfies (4.4), (4.5) and (4.6.). $\diamond$

Remark 4.12. When the domain Ω' is only Lipschitz-continuous, the vector potential $\vec{\psi}$ defined by Theorem 4.10 may not even belong to $W^1_k(\Omega)^3$, but it will always belong to $(X_N)^0_k(\Omega)$ (*cf.* an example with $k = -1$ in Girault, Giroire & Sequeira [16]). $\diamond$

5. The Stokes operator in $\mathbb{R}^3$

Let $\mathcal{J}$ denote the nonhomogeneous Stokes operator:

$$
(\vec{u}, p) \mapsto (-\nu\,\Delta\vec{u} + \nabla p, -\operatorname{div}\vec{u}),
$$

where ν is a given strictly positive constant. We need to characterize its null space. For this, we introduce the following subspace of $\mathbb{P}^3_k \times \mathbb{P}_{k-1}$:

$$
\mathcal{S}_k(\mathbb{R}^3) = \left\{ (\vec{r}, s) \in \mathbb{P}^3_k \times \mathcal{P}^\Delta_{k-1} \,;\, \operatorname{div}\vec{r} = 0 \text{ and } -\nu\,\Delta\vec{r} + \nabla s = \vec{0} \right\},
$$

with the convention that $\mathcal{S}_k(\mathbb{R}^3)$ is reduced to $(\vec{0}, 0)$ when k is negative and that $\mathcal{S}_0(\mathbb{R}^3) = \mathbb{R}^3 \times \{0\}$. Some useful properties of $\mathcal{S}_k(\mathbb{R}^3)$ are listed in Lemma 5.2; beforehand, we require the following result.

Lemma 5.1. *For all integers* $k \geq 0$, *we have*

$$
\mathcal{P}^\Delta_k = \operatorname{div}((\mathcal{P}^\Delta_{k+1})^3) .
$$

Proof : The case where $k = 0$ is trivial. Let us consider $k \geq 1$. Let p belong to $\mathcal{P}^\Delta_k$. Applying twice Lemma 2.4, there exists first $\vec{\mu}$ in $\mathbb{P}^3_k$ such that $\mathbf{curl}\,\vec{\mu} = \nabla p$ and $\operatorname{div}\vec{\mu} = 0$. And next, there exists $\vec{q}$ in $\mathbb{P}^3_{k+1}$ such that $\mathbf{curl}\,\vec{q} = \vec{\mu}$ and $\operatorname{div}\vec{q} = p$. Thus,

$$
\Delta\vec{q} = -\mathbf{curl}\,\mathbf{curl}\,\vec{q} + \nabla\operatorname{div}\vec{q} = \vec{0} . \qquad \diamond
$$

Lemma 5.2.

1. Let $\mathcal{G}_k$ be defined by (2.6). For all integers $k \geq 0$, we have:

$$\mathcal{G}_k \times \{0\} \subset \mathcal{S}_k(\mathbb{R}^3).$$

2. For all integers $k \geq 0$, each $\vec{p}$ in $\left(\mathcal{P}_k^\Delta\right)^3$ has the (nonunique) decomposition:

$$\vec{p} = \nabla q + \vec{w},$$

where q belongs to $\mathbb{P}_{k+1}$, $\vec{w}$ belongs to $\mathbb{P}_k^3$ and the pair $(\vec{w}, -\nu \Delta q)$ belongs to $\mathcal{S}_k(\mathbb{R}^3)$. Conversely, when p sweeps $\left(\mathcal{P}_k^\Delta\right)^3$ its associated pairs $(\vec{w}, -\nu \Delta q)$ sweep $\mathcal{S}_k(\mathbb{R}^3)$.

3. For all integers $k \geq 0$, for each s in $\mathcal{P}_{k-1}^\Delta$, there exists $\vec{r}$ in $\mathbb{P}_k^3$ such that the pair $(\vec{r}, s)$ belongs to $\mathcal{S}_k(\mathbb{R}^3)$.

4. For all integers m in $\mathbb{Z}$ and $k \geq 1$, the pair $(\vec{u}, p)$ in $V_{m-k}^{m+2}(\mathbb{R}^3) \times W_{m-k}^{m+1}(\mathbb{R}^3)$ is a solution of

$$-\nu \Delta \vec{u} + \nabla p = \vec{0} \quad , \quad \operatorname{div} \vec{u} = 0 \text{ in } \mathbb{R}^3 \tag{5.1}$$

if and only if the pair $(\vec{u}, p)$ belongs to $\mathcal{S}_k(\mathbb{R}^3)$. Thus $\mathcal{S}_k(\mathbb{R}^3)$ is the null space of the Stokes operator in $W_{m-k}^{m+2}(\mathbb{R}^3)^3 \times W_{m-k}^{m+1}(\mathbb{R}^3)$.

Proof : The first property is trivial.

When $k = 0$ the second property holds trivially with $\vec{w} = \vec{p}$ and $q = 0$. When $k \geq 1$, we solve $\Delta q = \operatorname{div} \vec{p}$, with q in $\mathbb{P}_{k+1}$, and set $\vec{w} = \vec{p} - \nabla q$. We easily check that $(\vec{w}, -\nu \Delta q)$ belongs to $\mathcal{S}_k(\mathbb{R}^3)$. The decomposition is obviously not unique, for if $\vec{p} = \vec{0}$, it can always be written $\vec{p} = \nabla q - \nabla q$, for any q in $\mathcal{P}_{k+1}^\Delta$.

Conversely, take any $(\vec{r}, s)$ in $\mathcal{S}_k(\mathbb{R}^3)$ and solve $-\nu \Delta q = s$, with q in $\mathbb{P}_{k+1}$. Then $\vec{r} + \nabla q$ belongs trivially to $\left(\mathcal{P}_k^\Delta\right)^3$.

The third property follows readily from Lemma 5.1 and the above decomposition. Indeed, if s belongs to $\mathcal{P}_{k-1}^\Delta$, then there exists $\vec{p}$ in $\left(\mathcal{P}_k^\Delta\right)^3$ such that $s = -\nu \operatorname{div} \vec{p}$. Then Property 2 implies that there exists q in $\mathbb{P}_{k+1}$ and $\vec{r}$ in $\mathbb{P}_k^3$ such that $\vec{p} = \nabla q + \vec{r}$, and the pair $(\vec{r}, -\nu \Delta q = s)$ belongs to $\mathcal{S}_k(\mathbb{R}^3)$.

For the fourth property, if $(\vec{u}, p)$ is a solution of (5.1) in $V_{m-k}^{m+2}(\mathbb{R}^3) \times W_{m-k}^{m+1}(\mathbb{R}^3)$ then $\Delta p = 0$ and therefore p belongs to $\mathcal{P}_{k-1}^\Delta$ and in turn the pair $(\vec{u}, p)$ belongs to $\mathcal{S}_k(\mathbb{R}^3)$. The converse is obvious. $\diamond$

Theorem 5.3. Let m belong to $\mathbb{Z}$ and $k \geq 1$ belong to $\mathbb{N}$. The Stokes operator $\mathcal{J}$ is an isomorphism from $W_{m+k}^{m+2}(\mathbb{R}^3)^3 \times W_{m+k}^{m+1}(\mathbb{R}^3)$ onto $\left[W_{m+k}^m(\mathbb{R}^3)^3 \times W_{m+k}^{m+1}(\mathbb{R}^3)\right] \perp \mathcal{S}_{k-2}(\mathbb{R}^3)$. In particular, when $k = 1$, the nonhomogeneous Stokes problem has a unique solution without any orthogonality constraint on the data.

Proof : Assume that the nonhomogeneous Stokes problem:

$$-\nu \Delta \vec{u} + \nabla p = \vec{f} \text{ in } \mathbb{R}^3 \quad , \quad -\operatorname{div} \vec{u} = g \text{ in } \mathbb{R}^3 \tag{5.2}$$

has a solution $\vec{u}$ in $W^{m+2}_{m+k}(\mathbb{R}^3)^3$ and p in $W^{m+1}_{m+k}(\mathbb{R}^3)$. Take the divergence of both sides of the first equation in (5.2):

$$\Delta p = \operatorname{div} \vec{f} + \nu \operatorname{div} \Delta \vec{u} = \operatorname{div} \vec{f} - \nu \Delta g \ . \tag{5.3}$$

Then, according to Theorem 1.4, $\operatorname{div} \vec{f} - \nu \Delta g$ must be orthogonal to $\mathcal{P}^\Delta_{k-1}$. On the other hand,

$$-\nu \Delta \vec{u} = \vec{f} - \nabla p$$

and it follows again from Theorem 1.4 that $\vec{f} - \nabla p$ must be orthogonal to $\left(\mathcal{P}^\Delta_{k-2}\right)^3$. When $k = 1$, these two orthogonalities are satisfied without any condition on $\vec{f}$ and g. This is consistent with the statement of the theorem since in this case, $\mathcal{S}_{k-2}(\mathbb{R}^3) = \{(\vec{0}, 0)\}$. Consider the case where $k \geq 2$; applying Property 2 of Lemma 5.2, each $\vec{z}$ in $\left(\mathcal{P}^\Delta_{k-2}\right)^3$ is of the form

$$\vec{z} = \nabla q + \vec{w},$$

with q in $\mathbb{P}_{k-1}$, $\vec{w}$ in $\mathbb{P}^3_{k-2}$ and $\operatorname{div} \vec{w} = 0$. Therefore, this last relationship implies that

$$0 = \langle \vec{f} - \nabla p, \vec{z} \rangle = \langle \vec{f}, \vec{w} \rangle - \langle \operatorname{div} \vec{f}, q \rangle + \langle \Delta p, q \rangle + \langle p, \operatorname{div} \vec{w} \rangle \ .$$

With (5.3), this gives

$$\langle \vec{f}, \vec{w} \rangle + \langle g, -\nu \Delta q \rangle = 0 \ .$$

But it follows from Lemma 5.2 that the pair $(\vec{w}, -\nu \Delta q)$ sweeps $\mathcal{S}_{k-2}(\mathbb{R}^3)$. Hence $(\vec{f}, g)$ must be orthogonal to $\mathcal{S}_{k-2}(\mathbb{R}^3)$. Note that this implies automatically that $\operatorname{div} \vec{f} - \nu \Delta g$ must be orthogonal to $\mathcal{P}^\Delta_{k-1}$, for if p belongs to $\mathcal{P}^\Delta_{k-1}$,

$$\langle \operatorname{div} \vec{f} - \nu \Delta g, p \rangle = -\langle \vec{f}, \nabla p \rangle - \langle \nu g, \Delta p \rangle = -\langle \vec{f}, \nabla p \rangle \ ,$$

and according to Property 1 of Lemma 5.2, the pair $(\nabla p, 0)$ belongs to $\mathcal{S}_{k-2}(\mathbb{R}^3)$.

Conversely, let the pair $(\vec{f}, g)$ be given in $\left[W^m_{m+k}(\mathbb{R}^3)^3 \times W^{m+1}_{m+k}(\mathbb{R}^3)\right] \perp \mathcal{S}_{k-2}(\mathbb{R}^3)$. Then, the above argument shows that, $\operatorname{div} \vec{f} - \nu \Delta g$ is orthogonal to $\mathcal{P}^\Delta_{k-1}$. Therefore Theorem 1.4 implies that there exists a unique p in $W^{m+1}_{m+k}(\mathbb{R}^3)$ such that (5.3) holds:

$$\Delta p = \operatorname{div} \vec{f} - \nu \Delta g \, ,$$

$$\|p\|_{W^{m+1}_{m+k}(\mathbb{R}^3)} \leq C_1 \|\operatorname{div} \vec{f} - \nu \Delta g\|_{W^{m-1}_{m+k}(\mathbb{R}^3)} \leq C_2 \left(\|\vec{f}\|_{W^m_{m+k}(\mathbb{R}^3)} + \|g\|_{W^{m+1}_{m+k}(\mathbb{R}^3)}\right) \, .$$

Similarly, we derive as above that $\vec{f} - \nabla p$ is orthogonal to $\left(\mathcal{P}^\Delta_{k-2}\right)^3$. Hence, another application of Theorem 1.4 shows that there exists a unique $\vec{u}$ in $W^{m+2}_{m+k}(\mathbb{R}^3)^3$ such that

$$-\nu \Delta \vec{u} = \vec{f} - \nabla p \, ,$$

$$\|\vec{u}\|_{W^{m+2}_{m+k}(\mathbb{R}^3)} \le C_3 \|\vec{f} - \nabla p\|_{W^m_{m+k}(\mathbb{R}^3)} \le C_4 \left(\|\vec{f}\|_{W^m_{m+k}(\mathbb{R}^3)} + \|g\|_{W^{m+1}_{m+k}(\mathbb{R}^3)}\right).$$

It remains to prove the second equation in (5.2). It stems from the first equation in (5.2) and (5.3) that

$$\Delta(\operatorname{div} \vec{u} + g) = 0.$$

In view of (1.6) and (1.4), this implies that $-\operatorname{div} \vec{u} = g$. ◇

Theorem 5.4. *Let the integers m and k belong to $\mathbb{Z}$ with $k \le 0$. The Stokes operator $\mathcal{J}$ is an isomorphism from $\left[W^{m+2}_{m+k}(\mathbb{R}^3)^3 \times W^{m+1}_{m+k}(\mathbb{R}^3)\right]/\mathcal{S}_{-k}(\mathbb{R}^3)$ onto $W^m_{m+k}(\mathbb{R}^3)^3 \times W^{m+1}_{m+k}(\mathbb{R}^3)$.*

Proof : The proof proceeds by a duality argument. For $\vec{f}$ in $W^m_{m+k}(\mathbb{R}^3)^3$ and g in $W^{m+1}_{m+k}(\mathbb{R}^3)$, the nonhomogeneous Stokes problem (5.2) is equivalent to the weak variational problem:

Find a pair $(\vec{u}, p)$ in $\left[W^{m+2}_{m+k}(\mathbb{R}^3)^3 \times W^{m+1}_{m+k}(\mathbb{R}^3)\right]/\mathcal{S}_{-k}(\mathbb{R}^3)$ such that $\forall (\vec{v}, q) \in W^{-m}_{-(m+k)}(\mathbb{R}^3)^3 \times W^{-m-1}_{-(m+k)}(\mathbb{R}^3)$

$$\langle \vec{u}, -\nu \Delta \vec{v} + \nabla q \rangle - \langle p, \operatorname{div} \vec{v} \rangle = \langle \vec{f}, \vec{v} \rangle + \langle g, q \rangle. \tag{5.4}$$

Indeed, if $(\vec{u}, p)$ is a solution of (5.2) in the above spaces, then by an easy density argument, we derive

$$\forall \vec{v} \in W^{-m}_{-(m+k)}(\mathbb{R}^3)^3 \ , \ \langle \vec{u}, -\nu \Delta \vec{v} \rangle - \langle p, \operatorname{div} \vec{v} \rangle = \langle \vec{f}, \vec{v} \rangle,$$

$$\forall q \in W^{-m-1}_{-(m+k)}(\mathbb{R}^3) \ , \ \langle \vec{u}, \nabla q \rangle = \langle g, q \rangle.$$

Thus, adding these two equations and observing that each pair $(\vec{r}, s)$ in $\mathcal{S}_{-k}(\mathbb{R}^3)$ satisfies

$$\forall (\vec{v}, q) \in W^{-m}_{-(m+k)}(\mathbb{R}^3)^3 \times W^{-m-1}_{-(m+k)}(\mathbb{R}^3) \ , \ \langle \vec{r}, -\nu \Delta \vec{v} + \nabla q \rangle - \langle s, \operatorname{div} \vec{v} \rangle = 0,$$

we obtain (5.4).

Conversely, if $(\vec{u}, p)$ is a solution of (5.4), then by taking first $\vec{v} = 0$ and next $q = 0$, we obtain on one hand

$$-\operatorname{div} \vec{u} = g,$$

and on the other hand

$$-\nu \Delta \vec{u} + \nabla p = \vec{f}.$$

Now, let us solve problem (5.4). We apply Theorem 5.3 and we denote by $\mathcal{J}^{-1}$ the inverse isomorphism of $\mathcal{J}$. Thus $\mathcal{J}^{-1}$ maps

$$\left[W^{-m-2}_{-(m+k)}(\mathbb{R}^3)^3 \times W^{-m-1}_{-(m+k)}(\mathbb{R}^3)\right] \perp \mathcal{S}_{-k}(\mathbb{R}^3) \ \text{ onto } \ W^{-m}_{-(m+k)}(\mathbb{R}^3)^3 \times W^{-m-1}_{-(m+k)}(\mathbb{R}^3)$$

. Hence problem (5.4) has the equivalent formulation:
Find a pair $(\vec{u}, p)$ in $\left[W^{m+2}_{m+k}(\mathbb{R}^3)^3 \times W^{m+1}_{m+k}(\mathbb{R}^3) \right] / \mathcal{S}_{-k}(\mathbb{R}^3)$ such that

$$\forall (\vec{z}, \zeta) \in \left[W^{-m-2}_{-(m+k)}(\mathbb{R}^3)^3 \times W^{-m-1}_{-(m+k)}(\mathbb{R}^3) \right] \perp \mathcal{S}_{-k}(\mathbb{R}^3),$$

$$\langle (\vec{u}, p), (\vec{z}, \zeta) \rangle = \langle (\vec{f}, g), \mathcal{J}^{-1}(\vec{z}, \zeta) \rangle. \tag{5.5}$$

Applying Riesz Representation Theorem, we see that problem (5.5) has a unique solution $(\vec{u}, p)$ in $\left[W^{m+2}_{m+k}(\mathbb{R}^3)^3 \times W^{m+1}_{m+k}(\mathbb{R}^3) \right] / \mathcal{S}_{-k}(\mathbb{R}^3)$ and

$$\inf_{(\vec{r}, s) \in \mathcal{S}_{-k}(\mathbb{R}^3)} \left(\| \vec{u} + \vec{r} \|_{W^{m+2}_{m+k}(\mathbb{R}^3)} + \| p + s \|_{W^{m+1}_{m+k}(\mathbb{R}^3)} \right)$$

$$\leq \| \mathcal{J}^{-1} \| (\| \vec{f} \|_{W^m_{m+k}(\mathbb{R}^3)} + \| g \|_{W^{m+1}_{m+k}(\mathbb{R}^3)}). \qquad \Diamond$$

The duality argument used in this proof has been introduced by Lions & Magenes in [26] (*cf.* also Nečas [28]) in the case of elliptic problems. It has been applied to the Stokes operator by Giga [10] and later by Amrouche & Girault [1,2].

6. The Stokes problem in exterior domains

The fully nonhomogeneous Stokes problem

$$\begin{aligned} -\nu \Delta \vec{u} + \nabla p &= \vec{f} \quad \text{in } \Omega, \\ -\operatorname{div} \vec{u} &= h \quad \text{in } \Omega, \\ \vec{u} &= \vec{g} \quad \text{on } \Gamma, \end{aligned} \tag{6.1}$$

has been solved by several authors in an unbounded domain. The result we shall use (and which can be found for instance in [9] and [15]) is:

Theorem 6.1. *Let the domain Ω' be Lipschitz-continuous. Then for all right-hand sides $\vec{f}$ in $W_0^{-1}(\Omega)^3$, h in $L^2(\Omega)$ and $\vec{g}$ in $H^{\frac{1}{2}}(\Gamma)^3$, the Stokes problem (6.1) has a unique solution $\vec{u}$ in $W_0^1(\Omega)^3$ and p in $L^2(\Omega)$ that depends continuously on the data.*

Furthermore, when the domain and the data are smoother, it can be shown that the solution has accordingly more regularity (*cf.* for instance [9], [19] or [15]). The purpose of this section is to extend the statement of this theorem to other exponents of the weight. Our approach is inspired by that used by Giroire in [17] for the Laplace operator: first calculate the kernel of the Stokes operator for negative values of the exponent, next show existence of solutions for negative exponents and from there derive by duality the existence of the (unique) solution for positive values of the exponent, provided the right-hand sides satisfy adequate orthogonality conditions.

We shall require the following space, related to the Stokes operator with "Neumann" boundary conditions. For all k in $\mathbb{Z}$, let

$$H_k(\mathcal{S};\Omega) = \{\vartheta = (\sigma, q) \in W_k^0(\Omega)^{3\times 3} \times W_k^0(\Omega);\ -\operatorname{div}\sigma + \nabla q \in W_{k+1}^0(\Omega)^3\}, \quad (6.2)$$

which is a Hilbert space for the norm

$$\|\vartheta\|_{H_k(\mathcal{S};\Omega)} = \left\{\|\sigma\|_{W_k^0(\Omega)}^2 + \|q\|_{W_k^0(\Omega)}^2 + \|-\operatorname{div}\sigma + \nabla q\|_{W_{k+1}^0(\Omega)}^2\right\}^{\frac{1}{2}}.$$

Our first lemma states a Green's formula on this space.

Lemma 6.2. *Assume that Ω' is Lipschitz-continuous. For all integers k in $\mathbb{Z}$, the product space $\mathcal{D}(\overline{\Omega})^{3\times 3} \times \mathcal{D}(\overline{\Omega})$ is dense in $H_k(\mathcal{S};\Omega)$, the trace mapping defined on $\mathcal{D}(\overline{\Omega})^{3\times 3} \times \mathcal{D}(\overline{\Omega})$:*

$$\vartheta \mapsto -\sigma \cdot \vec{n} + q\vec{n} \quad \text{on } \Gamma$$

can be extended by density from $H_k(\mathcal{S};\Omega)$ into $H^{-\frac{1}{2}}(\Gamma)^3$ and the following Green's formula holds for all ϑ in $H_k(\mathcal{S};\Omega)$ and for all $\vec{v}$ in $W_{-k}^1(\Omega)^3$:

$$\langle -\sigma \cdot \vec{n} + q\vec{n}, \vec{v}\rangle_\Gamma = \int_\Omega (-\operatorname{div}\sigma + \nabla q) \cdot \vec{v}\, d\mathbf{x} - \int_\Omega \sigma \cdot \nabla\vec{v}\, d\mathbf{x} + \int_\Omega q\operatorname{div}\vec{v}\, d\mathbf{x}. \quad (6.3)$$

Proof : We only sketch the proof, because its arguments are quite standard. First, the above density is valid on a bounded domain with the above spaces defined without weights. This can be readily proved on a star-shaped domain by the technique introduced by Temam in [33] and then extended to a Lipschitz-continuous domain, since it is the finite union of star-shaped domains. The Green's formula (6.3) is an easy consequence of this density.

Once the above density is proved for a bounded domain, it carries over to an exterior domain by the truncation technique introduced by Hanouzet in [21]. Then (6.3) follows again from this density. $\Diamond$

Now, we want to characterize the kernel of the Stokes operator. More precisely, for any integer $k \geq 1$, we wish to find all solution pairs $(\vec{v}, q)$ in $W_{-k}^1(\Omega)^3 \times W_{-k}^0(\Omega)$ of

$$\begin{aligned}
-\nu\Delta\vec{v} + \nabla q &= \vec{0} \quad \text{in } \Omega, \\
\operatorname{div}\vec{v} &= 0 \quad \text{in } \Omega, \\
\vec{v} &= \vec{0} \quad \text{on } \Gamma.
\end{aligned} \quad (6.4)$$

This space will be denoted by $\mathcal{S}_{k-1}(\Omega)$ and is an extension of $\mathcal{S}_{k-1}(\mathbb{R}^3)$.

Theorem 6.3. *Let Ω' be Lipschitz-continuous. For every integer $k \geq 1$, the kernel of the Stokes operator, $\mathcal{S}_{k-1}(\Omega)$, in $W^1_{-k}(\Omega)^3 \times W^0_{-k}(\Omega)$, is the space of all pairs $(\vec{v}, q)$ of the form:*

$$\vec{v} = \vec{w}(\vec{r}, s) - \vec{r} \quad , \quad q = z(\vec{r}, s) - s \,, \tag{6.5}$$

where the pair $(\vec{r}, s)$ sweeps the space $\mathcal{S}_{k-1}(\mathbb{R}^3)$ and the pair $(\vec{w}, z)$ is the unique solution in $W^1_0(\Omega)^3 \times L^2(\Omega)$ of the Stokes problem

$$
\begin{aligned}
-\nu\Delta\vec{w} + \nabla z &= \vec{0} && \text{in } \Omega \,, \\
\operatorname{div}\vec{w} &= 0 && \text{in } \Omega \,, \\
\vec{w} &= \vec{r} && \text{on } \Gamma \,.
\end{aligned}
\tag{6.6}
$$

When the domain Ω' is of class $\mathcal{C}^{m,1}$ for some integer $m \geq 1$, then $\mathcal{S}_{k-1}(\Omega)$ is contained in $W^{m+1}_{m-k}(\Omega)^3 \times W^m_{m-k}(\Omega)$.

Proof : Clearly, all elements of the form (6.5), (6.6) satisfy (6.4). Conversely, let the pair $(\vec{v}, q)$ be a solution of (6.4) in $W^1_{-k}(\Omega)^3 \times W^0_{-k}(\Omega)$ and let us prove that it is necessarily of the form (6.5), (6.6). To this end, we extend $\vec{v}$ and q by zero in Ω'; the extended pair, still denoted $(\vec{v}, q)$, belongs to $W^1_{-k}(\mathbb{R}^3)^3 \times W^0_{-k}(\mathbb{R}^3)$ and $\operatorname{div}\vec{v} = 0$ in $\mathbb{R}^3$. On one hand, let us calculate the distribution $-\nu\Delta\vec{v} + \nabla q$ in $\mathcal{D}'(\mathbb{R}^3)^3$. For any $\vec{\varphi}$ in $\mathcal{D}(\mathbb{R}^3)^3$, we have by definition

$$
\begin{aligned}
\langle -\nu\Delta\vec{v} + \nabla q, \vec{\varphi} \rangle &= -\nu\langle \vec{v}, \Delta\vec{\varphi} \rangle - \langle q, \operatorname{div}\vec{\varphi} \rangle \\
&= -\nu\int_\Omega \vec{v} \cdot \Delta\vec{\varphi}\, d\mathbf{x} - \int_\Omega q(\operatorname{div}\vec{\varphi})\, d\mathbf{x} \,.
\end{aligned}
$$

Then, as $\vec{\varphi}$ has a bounded support and $\vec{v}$ vanishes on Γ, two successive applications of Green's formula yield:

$$
\langle -\nu\Delta\vec{v} + \nabla q, \vec{\varphi} \rangle = \nu\int_\Omega \nabla\vec{v} \cdot \nabla\vec{\varphi}\, d\mathbf{x} - \int_\Omega q(\operatorname{div}\vec{\varphi})\, d\mathbf{x} = \langle \nu\frac{\partial\vec{v}}{\partial n} - q\vec{n}, \vec{\varphi} \rangle_\Gamma \,.
$$

Set $\vec{h} = \nu\frac{\partial\vec{v}}{\partial n} - q\vec{n}$ and observe that $\vec{h}$ belongs to $H^{-\frac{1}{2}}(\Gamma)^3$ because $(\nu\nabla\vec{v}, q)$ belongs to $H_{-k}(\mathcal{S}; \Omega)$. Thus denoting by $\vec{h}\delta_\Gamma$ the distribution

$$\forall\vec{\varphi} \in \mathcal{D}(\mathbb{R}^3)^3 \,,\ \langle \vec{h}\delta_\Gamma, \vec{\varphi} \rangle = \langle \vec{h}, \vec{\varphi} \rangle_\Gamma \,,$$

that belongs to $W^{-1}_0(\mathbb{R}^3)^3$, we can write

$$-\nu\Delta\vec{v} + \nabla q = \vec{h}\delta_\Gamma \,.$$

On the other hand, consider the problem:

Find $\vec{w}$ in $W_0^1(\mathbb{R}^3)^3$ and z in $L^2(\mathbb{R}^3)$ such that

$$-\nu\Delta\vec{w} + \nabla z = \vec{h}\delta_\Gamma \quad \text{and} \quad \text{div}\,\vec{w} = 0 \quad \text{in } \mathbb{R}^3.$$

As the right-hand side belongs to $W_0^{-1}(\mathbb{R}^3)^3$, this classical Stokes problem in $\mathbb{R}^3$ has a unique solution in the above spaces. Therefore, the difference pair $(\vec{r} = \vec{v} - \vec{w}, s = q - z)$ belongs to $W_{-k}^1(\mathbb{R}^3)^3 \times W_{-k}^0(\mathbb{R}^3)$ and satisfies

$$-\nu\Delta\vec{r} + \nabla s = \vec{0} \quad \text{and} \quad \text{div}\,\vec{r} = 0 \quad \text{in } \mathbb{R}^3.$$

Hence, it belongs to $\mathcal{S}_{k-1}(\mathbb{R}^3)$ and the pair $(\vec{v}, q)$ is of the form (6.5), (6.6).

Finally, if Ω' is of class $\mathcal{C}^{m,1}$, the regularity of $(\vec{v}, q)$ follows from the fact that $\vec{r}$ and s are polynomials and that the solution $(\vec{w}, z)$ of (6.6) belongs to $W_m^{m+1}(\Omega)^3 \times W_m^m(\Omega)$ (*cf.* [19]). This last result is also proved directly in Theorem 6.7 with $k = 0$. ◊

This theorem allows to solve the Stokes problem in $W_{-k}^1(\Omega)^3 \times W_{-k}^0(\Omega)$.

Theorem 6.4. *Let Ω' be Lipschitz-continuous. For any integer $k \geq 1$ and for all right-hand sides $\vec{f}$ in $W_{-k}^{-1}(\Omega)^3$, h in $W_{-k}^0(\Omega)$ and $\vec{g}$ in $H^{\frac{1}{2}}(\Gamma)^3$, the Stokes problem (6.1) has a unique solution pair $(\vec{u}, p)$ in $[W_{-k}^1(\Omega)^3 \times W_{-k}^0(\Omega)]/\mathcal{S}_{k-1}(\Omega)$ that depends continuously on the data:*

$$\inf_{(\vec{v},q)\in\mathcal{S}_{k-1}(\Omega)}(\|\vec{u} + \vec{v}\|_{W_{-k}^1(\Omega)} + \|p + q\|_{W_{-k}^0(\Omega)}) \leq$$

$$\leq C(\|\vec{f}\|_{W_{-k}^{-1}(\Omega)} + \|h\|_{W_{-k}^0(\Omega)} + \|\vec{g}\|_{H^{\frac{1}{2}}(\Gamma)}). \tag{6.7}$$

Proof : Let us construct one solution. First, we lift the divergence of $\vec{u}$. According to Lemma 3.1 there exists a unique $\vec{u}_h$ in $\overset{\circ}{W}{}_{-k}^1(\Omega)^3/V_{-k}^1(\Omega)$ such that $-\text{div}\,\vec{u}_h = h$ in Ω. Therefore, setting $\vec{u}_0 = \vec{u} - \vec{u}_h$ and $\vec{\ell} = \vec{f} + \nu\Delta\vec{u}_h$, problem (6.1) is equivalent to finding a pair $(\vec{u}_0, p)$ in $W_{-k}^1(\Omega)^3 \times W_{-k}^0(\Omega)$ such that

$$\begin{aligned}-\nu\Delta\vec{u}_0 + \nabla p &= \vec{\ell} \quad \text{in } \Omega, \\ \text{div}\,\vec{u}_0 &= 0 \quad \text{in } \Omega, \\ \vec{u}_0 &= \vec{g} \quad \text{on } \Gamma.\end{aligned} \tag{6.8}$$

Next, we extend $\vec{\ell}$ to Ω'. As it belongs to $W_{-k}^{-1}(\Omega)^3$, it has a (nonunique) decomposition of the form (*cf.* [17]):

$$\forall\vec{v} \in \overset{\circ}{W}{}_k^1(\Omega)^3, \quad \langle\vec{\ell}, \vec{v}\rangle = \int_\Omega \vec{\ell}_0 \cdot \vec{v}\,dx + \sum_{j=1}^3 \int_\Omega \vec{\ell}_j \cdot \partial_j\vec{v}\,dx,$$

where $\vec{\ell}_0$ belongs to $W^0_{-k+1}(\Omega)^3$ and the three $\vec{\ell}_j$ belong to $W^0_{-k}(\Omega)^3$. We extend $\vec{\ell}_0$ and $\vec{\ell}_j$ by zero in Ω'; the extended functions belong to $W^0_{-k+1}(\mathbb{R}^3)^3$ and $W^0_{-k}(\mathbb{R}^3)^3$ respectively and the corresponding distribution, that we still denote $\vec{\ell}$, belongs to $W^{-1}_{-k}(\mathbb{R}^3)^3$.

Then, we solve the Stokes problem in $\mathbb{R}^3$:

$$-\nu\Delta\vec{v} + \nabla q = \vec{\ell} \quad \text{and} \quad \operatorname{div}\vec{v} = 0 \quad \text{in } \mathbb{R}^3.$$

According to Theorem 5.4, this problem has a unique solution $(\vec{v}, q)$ in $[W^1_{-k}(\mathbb{R}^3)^3 \times W^0_{-k}(\mathbb{R}^3)]/\mathcal{S}_{k-1}(\mathbb{R}^3)$. Finally, we solve the Stokes problem in Ω:
Find $\vec{w}$ in $W^1_0(\Omega)^3$ and z in $L^2(\Omega)$ such that

$$\begin{aligned}
-\nu\Delta\vec{w} + \nabla z &= \vec{0} && \text{in } \Omega, \\
\operatorname{div}\vec{w} &= 0 && \text{in } \Omega, \\
\vec{w} &= \vec{g} - \vec{v} && \text{on } \Gamma,
\end{aligned}$$

which has a unique solution. Then the pair $(\vec{\theta} = \vec{v} + \vec{w}, \eta = q + z)$ belongs to $W^1_{-k}(\Omega)^3 \times W^0_{-k}(\Omega)$ and satisfies (6.8). This establishes existence of one solution for every right-hand side and obviously, the solution is unique up to an element of $\mathcal{S}_{k-1}(\Omega)$. The continuity is an easy consequence of this bijection and of the continuity of the inverse mapping. $\Diamond$

Now, we turn to the Stokes problem with positive exponents. First, we consider the problem with homogeneous boundary conditions:
Find $\vec{u}$ in $W^1_k(\Omega)^3$ and p in $W^0_k(\Omega)$ such that

$$\begin{aligned}
-\nu\Delta\vec{u} + \nabla p &= \vec{f} && \text{in } \Omega, \\
-\operatorname{div}\vec{u} &= h && \text{in } \Omega, \\
\vec{u} &= \vec{0} && \text{on } \Gamma,
\end{aligned} \tag{6.9}$$

with $\vec{f}$ given in $W^{-1}_k(\Omega)^3$ and h given in $W^0_k(\Omega)$ for strictly positive k. Since for negative values of k the null space of the Stokes operator is not reduced to zero, a solution for positive values of k will only exist under necessary orthogonality conditions on the right-hand sides. The next lemma states the precise result.

Lemma 6.5. *Let Ω' be Lipschitz-continuous. For any integer $k \geq 1$ and for all right-hand sides $\vec{f}$ in $W^{-1}_k(\Omega)^3$ and h in $W^0_k(\Omega)$, the Stokes problem (6.9) has a (unique) solution $(\vec{u}, p)$ in $W^1_k(\Omega)^3 \times W^0_k(\Omega)$ if and only if*

$$\forall(\vec{v}, q) \in \mathcal{S}_{k-1}(\Omega), \quad \langle\vec{f}, \vec{v}\rangle + \int_\Omega h\, q\, d\mathbf{x} = 0. \tag{6.10}$$

When it exists, the solution depends continuously on the data.

Proof : The uniqueness of the solution is obvious, so it suffices to prove existence. The proof proceeds by duality with respect to the weight. Let $(\vec{v}, q)$ be any pair in $\overset{\circ}{W}{}^{1}_{-k}(\Omega)^3 \times W^0_{-k}(\Omega)$ and let $(\vec{u}, p)$ be the solution of (6.9). Two applications of Green's formula yield:

$$\langle \vec{f}, \vec{v} \rangle = \langle -\nu \Delta \vec{u} + \nabla p, \vec{v} \rangle = \nu \int_\Omega \nabla \vec{u} \cdot \nabla \vec{v} \, d\mathbf{x} - \int_\Omega p (\operatorname{div} \vec{v}) \, d\mathbf{x}$$

$$= -\nu \langle \vec{u}, \Delta \vec{v} \rangle - \int_\Omega p (\operatorname{div} \vec{v}) \, d\mathbf{x}$$

$$\int_\Omega h \, q \, d\mathbf{x} = -\int_\Omega q (\operatorname{div} \vec{u}) \, d\mathbf{x} = \langle \vec{u}, \nabla q \rangle \, .$$

Thus adding these two equations we obtain

$$\langle \vec{u}, -\nu \Delta \vec{v} + \nabla q \rangle - \int_\Omega p (\operatorname{div} \vec{v}) \, d\mathbf{x} = \langle \vec{f}, \vec{v} \rangle + \int_\Omega h \, q \, d\mathbf{x} \tag{6.11}$$

$$\forall (\vec{v}, q) \in \overset{\circ}{W}{}^{1}_{-k}(\Omega)^3 \times W^0_{-k}(\Omega) \, .$$

Conversely, it is easy to check that each solution $(\vec{u}, p)$ in $\overset{\circ}{W}{}^{1}_{k}(\Omega)^3 \times W^0_k(\Omega)$ of (6.11) also satisfies (6.9). Hence, these two problems are equivalent and we shall solve (6.11).

First, problem (6.11) yields the necessary compatibility condition (6.10) by choosing for $(\vec{v}, q)$ all pairs in $\mathcal{S}_{k-1}(\Omega)$. Now, suppose that (6.10) is satisfied and let $(\vec{\ell}, e)$ be any pair in $W^{-1}_{-k}(\Omega)^3 \times W^0_{-k}(\Omega)$. According to Theorem 6.4, there exists a pair of functions $(\vec{v}, q)$ in $\overset{\circ}{W}{}^{1}_{-k}(\Omega)^3 \times W^0_{-k}(\Omega)$, unique up to an element of $\mathcal{S}_{k-1}(\Omega)$, such that

$$-\nu \Delta \vec{v} + \nabla q = \vec{\ell} \quad \text{and} \quad -\operatorname{div} \vec{v} = e \quad \text{in } \Omega \, . \tag{6.12}$$

Therefore (6.11) implies

$$\langle \vec{u}, \vec{\ell} \rangle + \int_\Omega p \, e \, d\mathbf{x} = \langle \vec{f}, \vec{v} \rangle + \int_\Omega h \, q \, d\mathbf{x} \, , \quad \forall (\vec{\ell}, e) \in W^{-1}_{-k}(\Omega)^3 \times W^0_{-k}(\Omega) \, , \tag{6.13}$$

where $(\vec{v}, q)$ is any solution in $\overset{\circ}{W}{}^{1}_{-k}(\Omega)^3 \times W^0_{-k}(\Omega)$ of the dual problem (6.12). As the converse is obvious, the two problems are also equivalent. Furthermore, by virtue of (6.10), the mapping

$$(\vec{\ell}, e) \mapsto \langle \vec{f}, \vec{v} \rangle + \int_\Omega h \, q \, d\mathbf{x}$$

is single-valued, linear and continuous:

$$|\langle \vec{f}, \vec{v} \rangle + \int_\Omega h\, q\, d\mathbf{x}| \le (\|\vec{f}\|_{W_k^{-1}(\Omega)} + \|h\|_{W_k^0(\Omega)})$$
$$\inf_{(\vec{r},s) \in \mathcal{S}_{k-1}(\Omega)} (\|\vec{v} + \vec{r}\|_{W_{-k}^1(\Omega)} + \|q + s\|_{W_{-k}^0(\Omega)})$$
$$\le C(\|\vec{f}\|_{W_k^{-1}(\Omega)} + \|h\|_{W_k^0(\Omega)})(\|\vec{\ell}\|_{W_{-k}^{-1}(\Omega)} + \|e\|_{W_{-k}^0(\Omega)}).$$

Then a straightforward application of Riesz Representation Theorem implies that (6.13) has a unique solution $(\vec{u}, p)$ in $\overset{\circ}{W}{}_k^1(\Omega)^3 \times W_k^0(\Omega)$. The continuity of the solution is straightforward. $\Diamond$

This lemma permits to solve the nonhomogeneous problem for strictly positive k: Find $\vec{u}$ in $W_k^1(\Omega)^3$ and p in $W_k^0(\Omega)$ satisfying (6.1):

$$-\nu\Delta\vec{u} + \nabla p = \vec{f} \quad \text{in } \Omega,$$
$$-\operatorname{div}\vec{u} = h \quad \text{in } \Omega,$$
$$\vec{u} = \vec{g} \quad \text{on } \Gamma,$$

with $\vec{f}$ given in $W_k^{-1}(\Omega)^3$, h given in $W_k^0(\Omega)$ and $\vec{g}$ given in $H^{\frac{1}{2}}(\Gamma)^3$.

Theorem 6.6. *Let Ω' be Lipschitz-continuous. For any integer $k \ge 1$ and for all right-hand sides $\vec{f}$ in $W_k^{-1}(\Omega)^3$, h in $W_k^0(\Omega)$ and $\vec{g}$ in $H^{\frac{1}{2}}(\Gamma)^3$, the Stokes problem (6.1) has a (unique) solution $(\vec{u}, p)$ in $W_k^1(\Omega)^3 \times W_k^0(\Omega)$ if and only if*

$$\forall(\vec{v}, q) \in \mathcal{S}_{k-1}(\Omega), \quad \langle \vec{f}, \vec{v} \rangle + \langle \vec{g}, -\nu\frac{\partial\vec{v}}{\partial n} + q\vec{n} \rangle_\Gamma + \int_\Omega h\, q\, d\mathbf{x} = 0. \qquad (6.14)$$

When it exists, the solution depends continuously on the data.

Proof : It suffices to lift $\vec{g}$: let $\vec{u}_g$ be any function in $W_k^1(\Omega)^3$ such that $\vec{u}_g = \vec{g}$ on Γ (we can take for instance a function with bounded support). Then setting $\vec{u}_0 = \vec{u} - \vec{u}_g$, $\vec{\ell} = \vec{f} + \nu\Delta\vec{u}_g$ and $e = h + \operatorname{div}\vec{u}_g$, problem (6.1) is equivalent to the homogeneous problem

Find $\vec{u}_0$ in $\overset{\circ}{W}{}_k^1(\Omega)^3$ and p in $W_k^0(\Omega)$ such that

$$-\nu\Delta\vec{u}_0 + \nabla p = \vec{\ell} \quad \text{and} \quad -\operatorname{div}\vec{u}_0 = e \quad \text{in } \Omega.$$

According to Lemma 6.5, this problem has a unique solution if and only if

$$\forall(\vec{v}, q) \in \mathcal{S}_{k-1}(\Omega), \quad \langle \vec{\ell}, \vec{v} \rangle + \int_\Omega e\, q\, d\mathbf{x} = 0.$$

Substituting the expressions of $\vec{\ell}$ and e yields

$$\langle \vec{f}, \vec{v} \rangle - \nu \int_\Omega \nabla \vec{u}_g \cdot \nabla \vec{v}\, dx + \int_\Omega h\, q\, dx + \int_\Omega q (\operatorname{div} \vec{u}_g)\, dx = 0\,.$$

But as the pair $(\nu \nabla \vec{v}, q)$ belongs to $H_{-k}(\mathcal{S}; \Omega)$ and $\vec{u}_g$ belongs to $W_k^1(\Omega)^3$, we can apply Green's formula (6.3) and we obtain precisely the condition (6.14).

Finally, it is easy to verify that the solution obtained is independent of the choice of $\vec{u}_g$ and depends continuously on the right-hand sides. $\diamond$

We end this section by studying the regularity of the solution of the Stokes problem when the boundary Γ and the right-hand sides have more regularity.

Theorem 6.7. *Let k belong to $\mathbb{Z}$ and m belong to $\mathbb{N}$. Suppose that Ω' is of class $\mathcal{C}^{m,1}$ and that the right-hand sides $\vec{f}$, h and $\vec{g}$ of the Stokes problem (6.1) belong respectively to $W_{m+k}^{m-1}(\Omega)^3$, $W_{m+k}^m(\Omega)$ and $H^{m+\frac{1}{2}}(\Gamma)^3$ and satisfy the compatibility condition (6.14) when $k \geq 1$. Then the solution of (6.1) belongs to $W_{m+k}^{m+1}(\Omega)^3 \times W_{m+k}^m(\Omega)$ or $[W_{m+k}^{m+1}(\Omega)^3 \times W_{m+k}^m(\Omega)]/\mathcal{S}_{-k-1}(\Omega)$ if $k \leq -1$, and depends continuously on the data.*

Proof: Existence of the solution has been proved in Theorems 6.4 and 6.6; it remains to establish the regularity. The proof proceeds by induction on m and is similar to that of Theorem 4.10 and Corollary 4.11. We already know that the result holds for $m = 0$; we assume that it is true for some $m \geq 0$ and show its validity for $m + 1$. To this end, we introduce the same partition of unity as in Theorem 4.10. With the same notations, we can write

$$\vec{u} = \lambda \vec{u} + \mu \vec{u} \quad , \quad p = \lambda p + \mu p\,,$$

where $(\vec{u}, p)$ satisfies (6.1). It suffices to examine $\lambda \vec{u}$ and λp in the bounded region Ω_{R_0+1} and $\mu \vec{u}$ and μp in the exterior region C_{R_0}.

After an easy calculation, we obtain that the pair $(\lambda \vec{u}, \lambda p)$ satisfies the following equations in Ω_{R_0+1}:

$$-\nu \Delta(\lambda \vec{u}) + \nabla(\lambda p) = \vec{\ell}_1 = \lambda \vec{f} + 2\nu \partial_j \mu \partial_j \vec{u} + \nu(\Delta \mu)\vec{u} - (\nabla \mu)p \quad \text{in } \Omega_{R_0+1}\,,$$

$$-\operatorname{div}(\lambda \vec{u}) = e_1 = \lambda h + \nabla \mu \cdot \vec{u} \quad \text{in } \Omega_{R_0+1}\,,$$

$$(\lambda \vec{u})_{|\Gamma} = \vec{g} \quad , \quad (\lambda \vec{u})_{|\partial B_{R_0+1}} = \vec{0}\,.$$

Note that the right-hand sides $\vec{\ell}_1$ and e_1 have indeed their supports in Ω_{R_0+1} and in view of the induction hypothesis, $\vec{\ell}_1$ belongs to $H^m(\Omega_{R_0+1})^3$ and e_1 belongs to $H^{m+1}(\Omega_{R_0+1})$. Then the regularity results for the Stokes problem in a bounded domain of class $\mathcal{C}^{m+1,1}$ (cf. [7], [9] or [1]) imply that $(\lambda \vec{u}, \lambda p)$ belongs to $H^{m+2}(\Omega_{R_0+1})^3 \times H^{m+1}(\Omega_{R_0+1})$.

Similarly, the pair $(\mu\vec{u}, \mu p)$ satisfies the following equations in C_{R_0}:

$$-\nu\Delta(\mu\vec{u}) + \nabla(\mu p) = \vec{\ell}_2 = \mu\vec{f} + 2\nu\partial_j\lambda\partial_j\vec{u} + \nu(\Delta\lambda)\vec{u} - (\nabla\lambda)p \quad \text{in } C_{R_0},$$

$$-\text{div}(\mu\vec{u}) = e_2 = \mu h + \nabla\lambda\cdot\vec{u} \quad \text{in } C_{R_0},$$

$$(\mu\vec{u})_{|\partial C_{R_0}} = \vec{0}.$$

As μ is very smooth and vanishes on B_{R_0}, the functions $\mu\vec{u}$ and μp can be extended by zero in B_{R_0} and the extended functions, still denoted $\mu\vec{u}$ and μp, belong respectively to $W^{m+1}_{m+k}(\mathbb{R}^3)^3$ and $W^m_{m+k}(\mathbb{R}^3)$. Similarly, owing to the supports of μ and λ and the induction hypothesis, $\vec{\ell}_2$ and e_2 can also be extended by zero in B_{R_0} and belong respectively to $W^m_{m+1+k}(\mathbb{R}^3)^3$ and $W^{m+1}_{m+1+k}(\mathbb{R}^3)$. Thus, we are led to study the regularity of a Stokes problem in $\mathbb{R}^3$. When $k \geq 1$, the right-hand sides $\vec{\ell}_2$ and e_2 satisfy the necessary compatibility condition in $\mathbb{R}^3$ (cf. Theorem 5.3) because we know that the solution exists in $W^1_k(\mathbb{R}^3)^3 \times W^0_k(\mathbb{R}^3)$. Hence, Theorem 5.3, if $k \geq 1$, or Theorem 5.4, if $k \leq 0$ imply that $(\mu\vec{u}, \mu p)$ belongs to $W^{m+2}_{m+1+k}(\mathbb{R}^3)^3 \times W^{m+1}_{m+1+k}(\mathbb{R}^3)$. Hence $(\vec{u}, p)$ belongs to $W^{m+2}_{m+1+k}(\Omega)^3 \times W^{m+1}_{m+1+k}(\Omega)$. The continuous dependence on the data follows easily from the existence and uniqueness of the solution and the continuity of the inverse mapping. $\diamond$

7. Some open problems

These lectures answer a few questions concerning steady flows past obstacles, but there are many open problems left to solve. Some are an easy exercise, like for example the Helmholtz decompositions of vector fields. Others are fairly straightforward, like the extension (at least of some of the results established here) to two-dimensional flows. The spaces are less easy to manipulate because, in the best of cases, they involve logarithmic weights. In the worst cases, they have a different definition; these are the X^m_k spaces defined and studied by Giroire in [17]. Nevertheless, their properties are similar to those of the spaces used here and it is reasonable to expect that the above arguments can be adapted to them. Along the same lines, is the extension to the $W^{m,p}_k$ spaces, where p means that the spaces are defined in terms of L^p instead of L^2. This extension is interesting in itself, but one of its most valuable applications is of course the Navier-Stokes equations.

The steady-state Navier-Stokes problem is a challenging but difficult problem, essentially because of its nonlinearity. The existence of a classical solution is well-known (cf. for example [23]), but what is not yet known is whether or not it can be put into a variational form (similar to that of the Stokes problem in $W^1_0(\Omega)^3 \times L^2(\Omega)$) from which existence and uniqueness of the solution can be deduced.

8. References

[1] Amrouche, C. and Girault, V., *Propriétés fonctionnelles d'opérateurs. Application au problème de Stokes en dimension quelconque*, Report LAN-UPMC 90025 (1990).

[2] Amrouche, C. and Girault, V., *Problèmes généralisés de Stokes*, Portugaliae Mathematica **49** (1992), no. 4, 463-503.

[3] Babuška, I., *The finite element method with Lagrangian multipliers*, Numer. Math. **20** (1973), 179-192.

[4] Borchers, W. and Sohr, H., *On the semigroup of the Stokes operator for exterior domains in L^q spaces*, Math. Z. **196** (1987), 415-425.

[5] Brezzi, F., *On the existence, uniqueness and approximation of saddle-point problems arising from Lagrange multipliers*, RAIRO – Anal. Num, R2 (1974), 129-151.

[6] Calderón, A.P., *Lebesgue space of differentiable functions and distributions'*, A.M.S., Providence, Proc. Symp. Pure Math., vol. 4, 1961., pp. 33-49.

[7] Cattabriga, L., *Su un problema al contorno relativo al sistema di equazioni di Stokes*, Sem. Mat. Univ. Padova **31** (1961), 308-340.

[8] Friedrichs, K.O., *Differential forms on Riemannian manifolds*, Comm. Pure Appl. Math. **8** (1955), 551-590.

[9] Galdi, G.P. and Simader, C.G., *Existence, uniqueness and L^q-estimates for the Stokes problem in an exterior domain*, Arch. Rat. Mech. Anal. **112** (1990), 291-318.

[10] Giga, Y.,, *Analyticity of the semigroup generated by the Stokes operator in L_r spaces*, Zeitschrift (1981), 297-329.

[11] Giga, Y. and Sohr, H. 'On the Stokes operator in exterior domains, J. Fac. Sci. Univ. Tokyo, Sect. IA **36** (1989), 103-130.

[12] Girault, V., *The gradient, divergence, curl and Stokes operators in weighted Sobolev spaces of $\mathbb{R}^3$*, J. Fac. Sci. Univ. Tokyo Sect. IA **39** (1992), no. 2, 279-307.

[13] Girault, V., *The Stokes problem and vector potential operator in three-dimensional exterior domains. An approach in weighted Sobolev spaces*, preprint LAN-UPMC (1992).

[14] Girault, V. and Raviart, P.A., *Finite Element Methods for Navier-Stokes Equations*, SCM **5**, Springer-Verlag, Berlin, 1986.

[15] Girault, V. and Sequeira, A., *A well-posed problem for the exterior Stokes equations in two and three dimensions*, Arch. Rat. Mech. and Anal. **114** (1991),

313-333.

[16] Girault, V., Giroire, J. and Sequeira, A., *A stream function-vorticity variational formulation for the exterior Stokes problem in weighted Sobolev spaces*, Math. Meth. in the App. Sciences **15** (1992), no. 5, 345-363.

[17] Giroire, J., *Etude de quelques problèmes aux limites extérieurs et résolution par équa-*
tions intégrales, Thèse UPMC, Paris, 1987.

[18] Grisvard, P., *Elliptic Problems in Non-Smooth Domains*, Monographs & Studies in Mathematics **24**, Pitman, 1985.

[19] Guirguis, G., *On the existence, uniqueness and regularity of the exterior Stokes problem in* $\mathbb{R}^3$, Com. in Partial Diff. Equat. **11** (1986), 567-594.

[20] Gunzburger, M., *Finite element methods for viscous incompressible flows*, Academic Press, London, 1989.

[21] Hanouzet, B., *Espaces de Sobolev avec poids. Application au problème de Dirichlet dans un demi-espace*, Rend. Sem. Univ. Padova **XLVI** (1971), 227-272.

[22] Héron, B., *Quelques propriétés des applications de traces dans des espaces de champs de vecteurs à divergence nulle*, Com. in Partial Diff. Equat. **6** (1981), no. 12, 1301-1334.

[23] Heywood, J.G., *The Navier-Stokes equations: On the existence, regularity and decay of solutions*, Indiana Univ. Math. J. **29** (1980), 639-681.

[24] Kozono, H. and Sohr, H., L^q*-regularity theory of the Stokes equations in exterior domains*, preprint.

[25] Ladyzhenskaya, O.A. and Solonnikov, V.A., *Some problems of vector analysis and generalized formulations of boundary-value problems for the Navier-Stokes equations*, J. of Soviet Math. **10** (1978), 257-286.

[26] Lions, J.-L. and Magenes, E., *Problemi ai limiti non homogenei (III)*, Ann. Scuola Norm. Sup. Pisa **15** (1961), 41-103.

[27] Miyakawa, T., *On nonstationary solutions of the Navier-Stokes equations in an exterior domain*, Hiroshima Math. J. **12** (1982), 111-140.

[28] Nečas, J., *Les Méthodes Directes en Théorie des Equations Elliptiques*, Masson, Paris, 1967.

[29] de Rham, G., *Variétés Differentiables*, Hermann, Paris, 1960.

[30] Schwarz, L., *Théorie des Distributions*, Hermann, Paris, 1966.

[31] Simon, J., *Primitives de distributions et applications*, or Distributions à valeurs vectorielles (to appear), Séminaire d'Analyse Clermont-Ferrand (1992).

[32] Tartar, L., *Topics in Nonlinear Analysis*, Publications Mathématiques d'Orsay,

Univ. Paris-Sud, 1978.

[33] Temam, R., *Navier-Stokes Equations. Theory and Analysis*, North-Holland, Amsterdam, 1985..

[34] von Wahl, W., *Lectures on the exterior problem for the Navier-Stokes equations*, Sonderforschungsbereich 256 "Nichtlineare partielle Differentialgleichungen", University of Bonn, 1989.

Vivette Girault
Analyse Numérique
Université Pierre et Marie Curie
Tour 55–65, 5ème étage, 4 place Jussieu
F–75252 Paris Cedex 05
France

WILLIAM G. LITVINOV
Models for Laminar and Turbulent Flows of Viscous and Nonlinear Viscous Fluids

Introduction

Theories of the flow of viscous and nonlinear viscous fluids are based on the Navier Stokes equations and on modifications of them (see [1]) for which the viscosity of the fluid depends on rates of strains (derivatives of the velocity). These models describe satisfactorily the slow laminar flows, but they are not suitable for the explanation of the flows with large gradients and turbulent flows.

Apparently, from the physical point of view the most general formulation for stationary and non-stationary problems is the mixed one, such that on one portion of the boundary the surface forces are prescribed, while on the other portion the velocities are given. For this formulation, existence results are known for Navier-Stokes equations (see [14] and references cited there) and for nonlinear viscous fluids under the neglect of nonlinear terms in the inertia forces [1]. There is also a statistical approach to the study of the Navier-Stokes equations (see [11] and references cited there).

We introduce and study models describing the stationary and non-stationary flows of the viscous and nonlinear viscous fluids for both the laminar and turbulent flows. In the case of turbulent flow these models describe the averaged velocities. The models include the main observed phenomena and give rise to correct mathematical problems in the sense of the existence of a solution for wide classes of data and for various formulations, including the mixed formulation.

1. Hydrodynamic experiments and Ladyzhenskaya equations

1.1. Viscous and nonlinear viscous fluids

Consider the simple shear flow. The fluid under consideration is held between the two parallel plates of infinite extent, the lower plate is fixed, the upper plate moves parallel to itself with constant speed v_0 (see Fig. 1). Then the velocity profile is $v(x) = v_0 x/h$, where h is the distance between the plates, x is the distance from the lower plate, Fig. 1.

Let τ denote the shearing stress, which is proportional to the force needed to move the upper plate, and let γ denote the shearing rate $\gamma = v_0/h$. According to the law of friction we have

$$\tau = \varphi\gamma, \qquad (1.1)$$

where φ is the viscosity of the fluid. For the Newtonian fluid $\varphi = \text{const} > 0$. But for many fluids τ is the nonlinear function of γ, i.e. $\varphi = \varphi(|\gamma|)$.

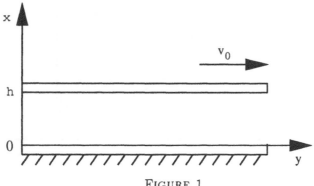

FIGURE 1.

Consider the power model, in this case $\varphi = a|\gamma|^{n-1}$, i.e.

$$\tau = a|\gamma|^{n-1}\gamma, \tag{1.2}$$

where a, n are positive constants. If $n = 1$, this is just the Newtonian fluid, if $n < 1$, then the viscosity decreases as the shearing rate is increased (dilatant fluid). As a rule with increasing of the shearing rate the structure of fluid is destroyed and the viscosity decreases [6], [7]. Therefore the most part of real fluids are pseudoplastic ones. Melted and dissolved polymers, oils, paints, emulsions, pastes, blood are examples of pseudoplastic fluids. Highly concentrated suspensions of solid particles are dilatant fluids.

For the power model (1.2) the problem of flow of fluid in the circular tube is exactly solved [8], and its solution is

$$v(r) = \frac{n}{n+1}\left|\frac{1}{2a}\frac{dp}{dz}\right|^{1/n} R^{\frac{n+1}{n}}\left[1 - \left(\frac{r}{R}\right)^{\frac{n+1}{n}}\right]. \tag{1.3}$$

Here $v(r)$ is the velocity of the fluid at a distance r from the axis of the tube, $\frac{dp}{dz} = \mathrm{const}$ – the drop of pressure per unit of the length of the tube, R is the radius of the tube.

The velocity profiles computed by formula (1.3) are shown in Fig. 2, line 1 is the profile of a Newtonian fluid, $n = 1$, line 2 a pseudoelastic fluid $n = 1/3$, line 3 a dilatant fluid $n = 2$, lines 4 a 5 are the limits as $n \to 0$ and $n \to \infty$, the corresponding profiles are rectangular and triangular ones.

1.2. Experimental results

In accordance with an experiment (see [6]) the relations between the mean velocity v_m in the circular tube $v_m = \left(2\int_0^R v(r)r\,dr\right)R^{-2}$ and the pressure p in the inflow

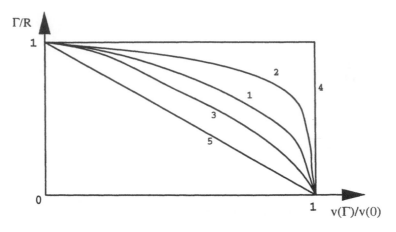

FIGURE 2.

of the tube have the forms shown in Fig. 3, line 1 a Newtonian fluid, line 2 a pseudoplastic fluid, a_1, a_2 being the points of transition to the turbulent flow. When the flow becomes turbulent there appear whirlwind flows and the relation p/v_m increases as v_m is increased, while for the pseudoplastic fluid in the range $[0, a_2]$ the relation p/v_m decreases as v_m is increased, and for the Newtonian fluid $p/v_m = \text{const}$ for all $v_m \in [0, a_1]$.

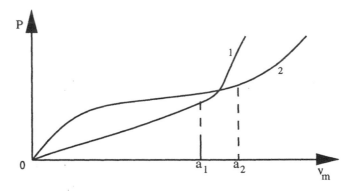

FIGURE 3.

The velocity profiles obtained by experiment for the flow of Newtonian fluid in the circular tube are shown in Fig. 4 (see [9]), lines 1,2,3 correspond to $v_m = b_1$, b_2, b_3. With this $b_1 < b_2 < b_3$, b_1 corresponds to laminar flow (line 1 is the parabola), lines 2 and 3 define the profiles of the averaged velocities for the turbulent flow.

The above experiments lead to the following conclusions.

1) For the turbulent flows of the Newtonian fluid in the circular tube the profiles of averaged velocities are analogous to the velocity profiles of pseudoplastic fluids for

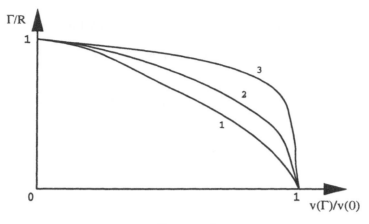

FIGURE 4.

the laminar flow (see Figures 2, 4), and $v_m/v(0) \to 1$ as $v_m \to \infty$, i.e. the velocity profile tends to rectangular one as $v_m \to \infty$.

2) As a rule the viscosity of real fluid is an nonincreasing function of the shearing rate. But the turbulent viscosity describing averaged velocities is an increasing function of the mean velocity v_m (see Fig. 3).

Since the profile of averaged velocities tends to rectangular one as $v_m \to \infty$, many authors consider a fluid for large v_m to be viscous in the boundary layer and unviscous outside of it. But such an approach is very rough because it does not take into account the main losses of energy, which take place outside the boundary layer due to whirlwind flows and velocity pulsations. In addition, there is usually no information about the thickness of the boundary layer, and some conditions on the boundary of the viscous and unviscous flows must be satisfied.

1.3. The Ladyzhenskaya model

In the case of arbitrary flow of incompressible nonlinear viscous fluid the shearing rate module $|\gamma|$ entered the expression $\varphi(|\gamma|)$ should be replaced by $|2I(v)|^{1/2}$, where $I(v)$ is the second invariant of the rate strain tensor

$$I(v) = \sum_{i,j=1}^{n} (\varepsilon_{ij}(v))^2, \qquad \varepsilon_{ij}(v) = \frac{1}{2}\left(\frac{\partial v_i}{\partial x_j} + \frac{\partial v_j}{\partial x_i}\right), \qquad (1.4)$$

and $v = (v_1, \ldots, v_n)$ is the velocity function. Then for the power model (1.2) the viscosity function gets the form

$$\varphi = a\,|2I(v)|^{(n-1)/2}. \qquad (1.5)$$

To describe the flow of viscous fluid for high velocity gradients Ladyzhenskaya has introduced two modifications of the Navier-Stokes equations [10], which define the flow of nonlinear viscous fluids with the following viscosity functions

$$\varphi = a_0 + a_1 I(v) \,, \tag{1.6}$$

$$\varphi = b_0 + b_1 \int_\Omega \sum_{i,j=1}^n \left(\frac{\partial v_i}{\partial x_j}(x,t) \right)^2 dx \,. \tag{1.7}$$

Here a_0, a_1, b_0, b_1 are positive constants, Ω is the domain in $\mathbb{R}^n$ in which the flow is studied, t is a time instant. The equations of motion of nonlinear viscous fluids are [1]

$$\rho \frac{\partial v_i}{\partial t} + \rho v_j \frac{\partial v_i}{\partial x_j} - 2 \frac{\partial}{\partial x_j} \left(\varphi \varepsilon_{ij}(v) \right) + \frac{\partial p}{\partial x_i} = f_i \ \text{ in } \ \Omega \times (0,T) \quad i = 1,\dots,n. \tag{1.8}$$

Here and below the summation over repeated indices is used, ρ is the density of the fluid, f_i are the components of the volume force function f, and p is the pressure.

Stationary and non-stationary problems for equations (1.8) with the functions φ defined by (1.6), (1.7) and with zero velocities on the boundary were studied in [2], [3], [10]. The solvability for stationary and non-stationary flows and the uniqueness in case of non-stationary flow were established. However, in case of the stationary flow of fluids with φ defined by (1.6) and (1.7) in the circular tube we have $v_m/v(0) < 2$, and $v_m/v(0) = 2$, respectively, for all v_m, while for the real fluids $v_m/v(0) \to 1$ as $v \to \infty$ (see Fig. 4).

We will now formulate and investigate the mathematical models which describe the phenomena shown in Figures 2–4.

2. Stationary problem

2.1. Formulation of the problem

We consider the stationary flow of the fluid in a bounded domain $\Omega \subset \mathbb{R}^n$ ($n = 2, 3$) with a Lipschitz continuous boundary S. Define a constitutive equation by

$$\sigma_{ij}(p,v) = -p\delta_{ij} + 2\varphi \left(I(v), \|v\|_4 \right) \varepsilon_{ij}(v) \,. \tag{2.1}$$

Here $\sigma_{ij}(p,v)$ are the components of the stress tensor depending on the pressure p and velocity $v = (v_1,\dots,v_n)$, δ_{ij} is the Kronecker symbol, φ is a viscosity function depending on $I(v)$ and $\|v\|_4$,

$$\|v\|_m = \left(\int_\Omega |v|^m \, dx \right)^{1/m}, \ m \in [1,\infty) \qquad |v| = \left(\sum_{i=1}^n v_i^2 \right)^{1/2}, \tag{2.2}$$

$I(v), \varepsilon_{ij}(v)$ being determined by (1.4).

82

It should be mentioned that we consider our problem in the immovable Euler coordinate system. Because the norm $\|v\| - m$ is defined by the length (module) $|v|$ of the vector v, and the length of a vector is independent of a coordinate system, the equation (2.1) is invariant for an arbitrary immovable Euler coordinate system. But in case of Galilean transformations, corresponding to the uniform motion (without an acceleration) of a coordinate system, equation (2.1) is non-invariant due to the term $\|v\|_4$.

Let S_1, S_2 be open non-empty subsets in S such that $S = \overline{S}_1 \cup \overline{S}_2$, and $S_1 \cap S_2$ is an empty set. Consider the problem:

find a pair of functions p, v satisfying

$$\rho v_j \frac{\partial v_i}{\partial x_j} - \frac{\partial \sigma_{ij}(p, v)}{\partial x_j} = K_j \qquad in\ \Omega \quad i = 1, ..., n, \tag{2.3}$$

$$\operatorname{div} v = 0 \ in\ \Omega \tag{2.4}$$

$$v\big|_{S_1} = 0, \qquad a_{ij}(p, v)v_j\big|_{S_2} = F_i \qquad i = 1, ..., n. \tag{2.5}$$

Here ρ is the density of fluid, $\rho = \mathrm{const} > 0$, K_i, F_i are the components of vectors of volume and surface forces $K = (K_1, ..., K_n)$, $F = (F_1, ..., F_n)$, v_j are the components of the unit outward normal v to S.

We suppose that φ is a continuous function on $\mathbb{R}_+^2$, where $\mathbb{R}_+ = \{z \in \mathbb{R}, z \geq 0\}$, satisfying the conditions

$$a_2 \geq \varphi(y_1, y_2) \geq a_1 \qquad \forall\, (y_1, y_2) \in \mathbb{R}_+ \times [0, a_0], \tag{2.6}$$

$$a_4 y_2^2 \geq \varphi(y_1, y_2) \geq a_3 y_2^2 \qquad \forall\, (y_1, y_2) \in \mathbb{R}_+ \times (a_0, \infty), \tag{2.7}$$

$$[\varphi(z_1^2, y_2)z_1 - \varphi(z_2^2, y_2)z_2](z_1 - z_2) \geq a_5(z_1 - z_2)^2 \quad \forall\, (z_1, z_2, y_2) \in \mathbb{R}_+^3, \tag{2.8}$$

where $a_0, ..., a_5$ are positive constants.

We now dwell on physical meanings of inequalities (2.6)–(2.8). The expressions (2.6), (2.7) designate that the viscosity function should be positive, and when $\|v\|_4 \geq a_0$ the viscosity must increase with the increase of $\|v\|_4$, a_0 being a point of the loss of stability. When $\|v\|_4 > a_0$, there appear whirlwind flows and pulsations of velocities, and so the viscosity of averaged velocities increases with the increase of $\|v\|_4$. This fact conforms to observations of flow patterns, see Fig. 3. The inequality (2.8) designates that for an arbitrary fixed value of $\|v\|_4$ the shear stress must increase with the increase of the shear rate (see [1]). The conditions (2.6)–(2.8) are natural from the physical point of view. We can in particular consider $\varphi(y_1, y_2) = \mathrm{const} \qquad \forall\, (y_1, y_2) \in \mathbb{R}_+ \times [0, a_0]$, where a_0 is an arbitrary positive constant. Therefore our model can describe all what is described by the Navier-Stokes equations. But when there does not exist a solution of the Navier-Stokes equations, the term $\|v\|_4$ in (2.1) comes forward as a regularizer and there exists a solution of our problem (see Theorem 3.1). For fixed values $y_2^{(i)} = \|v^{(i)}\|_4$, $i = 1, \ldots, k$ one can by experiment define the functions $f_i : y_1 \to \varphi(y_1, y_2^{(i)})$, and then define φ on $\mathbb{R}_+ \times \mathbb{R}_+$ by the interpolation and prolongation. Thus our model includes the observed phenomena.

We define the spaces X and V as follows:

$$X = \left\{ u \in H^1(\Omega)^n, u\big|_{S_1} = 0 \right\}, \tag{2.9}$$

$$V = \left\{ u \in H^1(\Omega)^n, \operatorname{div} u = 0, u\big|_{S_1} = 0 \right\}. \tag{2.10}$$

By X^*, V^* we denote the duals of X and V, respectively. In virtue of Korn's inequality the expression

$$\|u\| = \left(\int_\Omega I(u) \, dx \right)^{1/2}, \tag{2.11}$$

where $I(u)$ is defined by (1.4), determines the norm in X and V, which is equivalent to the norm of the space $H^1(\Omega)^n$; X and V being Hilbert spaces with the scalar product $(u, h)_X = \int_\Omega \varepsilon_{ij}(h) \, dx$. We now define an operator $L : X \to X^*$ by

$$(L(u), h) = 2 \int_\Omega \varphi(I(u), \|u\|_4) \varepsilon_{ij}(h) \, dx + \rho \int_\Omega u_j \frac{\partial u_i}{\partial x_j} h_i \, dx \qquad u, h \in X. \tag{2.12}$$

Consider the problem: *find $v \in V$ satisfying*

$$(L(v), h) = (K + F, h) \qquad \forall\, h \in V, \tag{2.13}$$

where

$$(K + F, h) = \int_\Omega K_i h_i \, dx + \int_{S_2} F_i h_i \, ds. \tag{2.14}$$

We suppose that

$$K, F \in X^*. \tag{2.15}$$

By the Green formula it is easy to see that if p, v is a smooth solution of problem (2.3)–(2.5) then v is a solution of problem (2.13). The inverse result is the following.

Theorem 2.1. *Let φ be a continuous function on $\mathbb{R}_+^2$ satisfying (2.6)–(2.8) and (2.15) holds. Let v be a solution of problem (2.13). Then there exists the unique function $p \in L_2(\Omega)$ such that (p, v) is a solution of problem (2.3)–(2.5) in the sense of distributions.*

For the proof of Theorem 2.1 we need the following lemma.

84

Lemma 2.1. *The operator* div *is an isomorphism from* $V^\perp$ *onto* $L_2(\Omega)$, *where* $V^\perp$ *is the orthogonal complement of* V *in* X.

PROOF : It is known that for an arbitrary $h \in L_2(\Omega)$ such that $\int_\Omega h \, dx = 0$ there exists a function $u \in \overset{o}{H}{}^1(\Omega)^n$ for which $\operatorname{div} u = h$ (see [12], [13]), and so by the Banach theorem it suffices to show the existence of a function w such that

$$w \in X, \qquad \operatorname{div} w = 1. \tag{2.16}$$

Take a function g satisfying

$$g \in H^1(\Omega)^n, \qquad \operatorname{supp} g \subset \Omega \cap S_2, \qquad \int_{S_2} g_i \nu_i \, ds = a \neq 0. \tag{2.17}$$

By the Green formula we obtain

$$\int_\Omega \left(\frac{\operatorname{mes} \Omega}{a} \operatorname{div} g - 1 \right) dx = 0, \tag{2.18}$$

and so for the function $h = \frac{\operatorname{mes} \Omega}{a} \operatorname{div} g - 1$ we have $h \in L_2(\Omega)$, $\int_\Omega h \, dx = 0$. Therefore there exists a function u such that $u \in H^{01}(\Omega)^n, \operatorname{div} u = h$, and so the function $w = \frac{\operatorname{mes} \Omega}{a} g - u$ satisfies (2.16). ∎

PROOF of the Theorem 2.1 : Let v be a solution of problem (2.13). By the embedding theorem we have $V \subset L_4(\Omega)^n$ for $n \leq 3$, and so $\sum\limits_{i,j=1} v_j \frac{\partial v_i}{\partial x_j} \in X*$, and by (2.6), (2.7) we get $\varphi(I(v), \|v\|_4) \in L_\infty(\Omega)$. Therefore

$$L(v) - K - F \in V^0,$$
$$V^0 = \{ z \in X^*, (z, h) = 0, \forall h \in V \}. \tag{2.19}$$

The space V^0 can be identified isometrically with $(V^\perp)^*$. Indeed, for $v \in X$ let $v^\perp$ denote the orthogonal projection of v on $V^\perp$. Then with $g \in (V^\perp)^*$ we associate an element $\tilde{g}$ on X^* defined by $(\tilde{g}, v) = (g, v^\perp \; \forall v \in X$. This permits to identify $(V^\perp)^*$ and V^0.

Let $\operatorname{div}^* \in \mathcal{L}(L_2(\Omega), (V^\perp)^*)$ be the adjoint operator of div , i.e.$(\operatorname{div} u, h) = (u, \operatorname{div}^* h) \; \forall u \in V^\perp, \; \forall h \in L_2(\Omega)$. It follows from Lemma 2.1 that div^* is an isomorphism from $L_2(\Omega)$ onto $(V^\perp)^*$, and so by (2.19) there exists the unique function $p \in L_2(\Omega)$ such that

$$(L(v) - K - F, h) = \int_\Omega p \operatorname{div} h \, dx \qquad \forall h \in X. \tag{2.20}$$

Taking here $h \in D(\Omega)^n$ we obtain (2.3), and taking then $h \in X$ we get the second equality of (2.5) ∎

2.2 Auxiliary statements

For an arbitrary fixed $u \in V$ we define a mapping $A_u : V \to V^*$ by

$$(A_u(v), h) = 2 \int_\Omega \varphi(I(v), \|u\|_4) \varepsilon_{ij}(v) \varepsilon_{ij}(h) \, dx, \qquad v, h \in V. \qquad (2.21)$$

2.2 Lemma. *Let the conditions (2.6)–(2.8) be satisfied and the operator A_u be defined by (2.21), where $u \in V$. Then A_u is a continuous mapping from V into V^*, and*

$$(A_u(v) - A_u(w), v - w) \ge 2a_5 \int_\Omega [I(v)^{1/2}]^2 \, dx \qquad \forall\, u, vw \in V, \qquad (2.22)$$

and from the condition $(A_u(v) - A_u(w), v - w) = 0$ it follows that $v = w$.

PROOF : Taking into account (2.21) and the inequality

$$|\varepsilon_{ij}(v)\varepsilon_{ij}(w)| \le I(v)^{1/2} I(w)^{1/2}, \qquad (2.23)$$

we obtain

$$(A_u(v) - A_u(w), v - w) = 2 \int_\Omega [\varphi(I(v), \|u\|_4) I(v) + \varphi(I(w), \|u\|_4) I(w) -$$
$$- \varphi(I(v), \|u\|_4) \varepsilon_{ij}(v) \varepsilon_{ij}(w) - \varphi(I(w), \|u\|_4) \varepsilon_{ij}(w) \varepsilon_{ij}(v)] \, dx \ge$$
$$\ge 2 \int_\Omega \left[\varphi(I(v), \|u\|_4) I(v)^{1/2} - \varphi(I(w), \|u\|_4) I(w)^{1/2} \right] \left[I(v)^{1/2} - I(w)^{1/2} \right] \, dx.$$

From here by (2.8) we obtain (2.22). Let now

$$(A_u(v) - A_u(w), v - w) = 0. \qquad (2.24)$$

Then it follows from (2.22) that $\varphi(I(v), \|u\|_4) = \varphi(I(w), \|u\|_4)$ almost everywhere in Ω, and taking into account (2.6), (2.7), (2.11), (2.24) we obtain $\|v - w\| = 0$.

It remains to show that A_u is a continuous mapping from V into V^*. Let $v_k \to v_0$. Then from the sequence $\{v_k\}$ we may choose a subsequence $\{v_m\}$ such that

$$I(v_m) \to I(v_0) \quad almost\ everywhere\ \text{in } \Omega. \qquad (2.25)$$

We take the following notations

$$I_m = I(v_m), \quad \varepsilon_{ijm} = \varepsilon_{ij}(v_m), \quad \varphi_m = \varphi(I(v_m), \|u\|_4) \quad m = 0, 1, 2, \ldots \qquad (2.26)$$

Taking into account (2.21) and the inequalities (2.6), (2.7) we get

$$
\begin{aligned}
|(A_u(v_m) - A_u(v_0), h)| &= \\
&= 2| \int_\Omega [\varphi_m(\varepsilon_{ijm} - \varepsilon_{ij0}) + (\varphi_m - \varphi_0)\varepsilon_{ij0}]\varepsilon_{ij}(h)\, dx| \le \\
&\le \left[c\|v_m - v_0\| + 2 \left(\int_\Omega (\varphi_m - \varphi_0)^2 I_0\, dx \right)^{\frac{1}{2}} \right] \|h\|.
\end{aligned}
\tag{2.27}
$$

By (2.6), (2.7), (2.25) we can apply the Lebesgue's theorem and obtain $\int_\Omega (\varphi_m - \varphi_0)^2 I_0\, dx \to 0$. Therefore

$$
A_u(v_m) \to A_u(v_0) \text{ in } V^*.
\tag{2.28}
$$

From an arbitrary subsequence $\{v_l\}$, chosen from $\{v_k\}$, we may thus choose a subsequence $\{v_m\}$ for which (2.28) holds. Therefore $A_u(v_k) \to A_u(v_0)$ in V^*. ∎

We define operators $L_i : V \to V^*$, $i = 1, 2$ by

$$
(L_1(u), h) = 2 \int_\Omega \varphi(I(u), \|u\|_4)\varepsilon_{ij}(u)\varepsilon_{ij}(h)\, dx,
\tag{2.29}
$$

$$
(L_2(u), h) = \rho \int_\Omega u_j \frac{\partial u_i}{\partial x_j} h_j\, dx.
\tag{2.30}
$$

It follows from (2.12), (2.21), (2.29), (2.30) that

$$
L(u) = L_1(u) + L_2(u),
\tag{2.31}
$$

$$
A_u(u) = L_1(u).
\tag{2.32}
$$

2.3 Lemma. *Suppose that*

$$
v_m \to v_0 \quad \text{weakly in } V.
\tag{2.33}
$$

Then $L_2(v_m) \to L_2(v_0)$ *strongly in* V^*.

PROOF : Denoting

$$
A_m = \rho \int_\Omega v_{0j} \left(\frac{\partial v_{mi}}{\partial x_j} - \frac{\partial v_{0i}}{\partial x_j} \right) h_i\, dx,
$$

and using the Hölder inequality and embedding theorems we obtain

$$
\begin{aligned}
|(L_2(v_m) - L_2(v_0), h)| &\le |\rho \int_\Omega (v_{mj} - v_{0j}) \frac{\partial v_{mi}}{\partial x_j} h_i\, dx| + |A_m| \le \\
&\le c\|v_m - v_0\|_4 \|v_m\| \|h\| + |A_m|.
\end{aligned}
\tag{2.34}
$$

Taking (2.10) into account and applying the Green formula we have

$$|A_m| \leq \rho | \int_\Omega v_{0j}(v_{mi} - v_{0i})\frac{\partial h_i}{\partial x_j}\, dx| + \rho | \int_{S_2} v_{0j}(v_{mi} - v_{0i})h_i\nu_j\, ds| \leq$$

$$\leq c\|v_0\|\,\|v_m - v_0\|_4\|h\| + \rho \sum_{i,j=1}^n \|v_{0j}\|_{L_4(S_2)}\|v_{mi} - v_{0i}\|_{L_2(S_2)}\|h_i\|_{L_4(S_2)}.$$

(2.35)

It follows from (2.33) that $v_m \to v_0$ in $L_4(\Omega)^n$, and $v_m \to v_0$ in $L_2(S_2)^n$, so (2.34), (2.35) give $L_2(v_m) \to L_2(v_0)$ in V^*. ∎

3. The existence theorem

Let $\{V_n\}$ be a sequence of finite dimensional subspaces of V such that

$$\lim_{n\to\infty} \inf_{z\in V_n} \|u - z\| = 0 \qquad \forall\, u \in V,$$

(3.1)

$$V_n \subset V_{n+1}.$$

(3.2)

We seek an approximate solution of problem (2.13) of the form

$$v_n \in V_n \qquad (L(v_n), h) = (K + F, h) \qquad \forall\, u \in V_n.$$

(3.3)

Theorem 3.1. *Suppose the conditions (2.6)–(2.8), (2.15), (3.1) and (3.2) are satisfied. Then there exists a solution v of problem (2.13), and for an arbitrary n there exists a solution v_n of problem (3.3). A subsequence $\{v_m\}$ such that $v_m \to v$ in V.*

PROOF :

1. For an arbitrary $u \in V$ taking into account (2.6), (2.7), (2.12), (2.15) we get

$$(L(u), u) - (K + F, u) \geq$$

$$\geq \begin{cases} 2a_1\|u\|^2 - (c + \|K + F\|_{V^*})\|u\| & \text{if } \|u\|_4 \leq a_0 \\ \|u\|_4^2\|u\|(2a_3\|u\| - c_1) - \|u\|\|K + F\|_{V^*} & \text{if } \|u\|_4 > a_0 \end{cases}$$

(3.4)

The lower expression of the right–hand side of (3.4) is not less than the value

$$\|u\|_4^2\|u\|(a_3\|u\| - c_1) + \|u\|(a_3 a_0^2\|u\| - \|K + F\|_{V^*}).$$

From here and (3.4) it follows that

$$(L(u), u) - (K + F, u) \geq 0 \quad \text{if} \quad \|u\| \geq r,$$

(3.5)

where

$$r = \max\left(\frac{c + \|K + F\|_{V^*}}{2a_1}, \frac{c_1}{a_3}, \frac{\|K + F\|_{V^*}}{a_3 a_0^2}\right).$$

Therefore, for an arbitrary n there exists a solution of problem (3.3) satisfying

$$\|v_n\| \leq r \qquad \forall\, n\,, \tag{3.6}$$

and so we can choose a subsequence $\{v_m\}$ such that

$$v_m \rightharpoonup v_0 \qquad \text{weakly in } V \tag{3.7}$$

$$v_m \to v_0 \qquad \text{in } L_4(\Omega)^n\,. \tag{3.8}$$

By (2.6), (2.7) and (3.6) we have $\|L_1(v_m)\|_{V^*} \leq C\ \forall\, m$, and we may extract a subsequence still denoted by $\{v_m\}$ for which

$$L_1(v_m) \rightharpoonup \chi \qquad \text{weakly in } V^*. \tag{3.9}$$

Let m_0 be a fixed positive integer and $h \in V_{m_0}$. By (2.31), (3.7), (3.9) and Lemma 2.3 we pass to the limit in (3.3) under n changed for m, then we obtain

$$(\chi, h) + (L_2(v_0), h) = (K + F, h) \quad \forall\, h \in V_{m_0}\,. \tag{3.10}$$

Here m_0 is an arbitrary positive integer and due to (3.1) we get

$$\chi + L_2(v_0) = K + F\,. \tag{3.11}$$

For $m = 0, 1, \ldots$ we define the mappings $J_m : V \mapsto V^*$ as follows

$$(J_m(u), h) = 2 \int_{\Omega} \varphi(I(u), \|v_m\|_4) \varepsilon_{ij}(u)\, \varepsilon_{ij}(h)\, dx\,, \quad u, h \in V\,. \tag{3.12}$$

It follows from (2.29) and (3.12) that

$$L_1(v_m) = J_m(v_m)\,, \tag{3.13}$$

and due to Lemma 2.1 we have

$$(J_m(v_m) - J_m(u), v_m - u) \geq 0 \quad \forall\, u \in V\,, \quad \forall\, m\,. \tag{3.14}$$

Taking (2.6), (2.7), (3.8) into account and applying the Lebesgue theorem we get

$$\varphi(I(u), \|v_m\|_4)\varepsilon_{ij}(u) \to \varphi(I(u), \|v_0\|_4)\varepsilon_{ij}(u) \text{ in } L_2(\Omega) \text{ as } m \to \infty\,. \tag{3.15}$$

The relations (3.7), (3.12) and (3.15) yield

$$\lim(J_m(u), u) = (J_0(u), u)\,, \tag{3.16}$$

$$\lim(J_m(u), v_m) = (J_0(u), v_0)\,, \tag{3.17}$$

By (3.9), (3.11), (3.13) we have

$$\lim \left(J_m(v_m), u \right) + (L_2(v_0), u) = (K + F, u), \tag{3.18}$$

and due to (3.3), (3.7), (3.17) and Lemma 2.3 we get

$$(J_m(v_m), v_m) = (K + F, v_m) - (L_2(v_m), v_m) \to (K + F, v_0) - (L_2(v_0), v_0). \tag{3.19}$$

Now by (3.16)–(3.19), passing to the limit in (3.14) we have

$$(K + F - L_2(v_0) - J_0(u), v_0 - u) \geq 0 \quad \forall\, u \in V. \tag{3.20}$$

Take here $u = v_0 - \gamma h$, $\gamma > 0$, $h \in V$ and let γ tend to zero. As J_0 is a continuous mapping (Lemma 2.2) we obtain $(K + F - L_2(v_0) - J_0(v_0), h) \geq 0 \quad \forall\, h \in V$. From here and (3.13) it follows that the function $v = v_0$ is a solution of problem (2.13).

2. We now show that

$$v_m \to v_0 \quad \text{strongly in } V. \tag{3.21}$$

By (2.11) we have

$$\|v_m - v_0\|^2 = \int\limits_\Omega I(v_m - v_0)\, dx = \int\limits_\Omega \left[I(v_m) - 2\varepsilon_{ij}(v_m)\,\varepsilon_{ij}(v_0) + I(v_0) \right] dx. \tag{3.22}$$

Due to (3.7) and (3.22) in order to prove (3.21) it is sufficient to show that

$$\lim \int\limits_\Omega I(v_m)\, dx = \int\limits_\Omega I(v_0)\, dx. \tag{3.23}$$

Let us denote

$$X_m = (J_m(v_m) - J_0(v_0), v_m - v_0). \tag{3.24}$$

It is obvious that

$$X_m = (J_m(v_m) - J_m(v_0), v_m - v_0) + (J_m(v_0) - J_0(v_0), v_m - v_0). \tag{3.25}$$

It follows from (3.7), (3.18), (3.19), (3.24) that $X_m \to 0$, the second addend in (3.25) tends also to zero, and so

$$(J_m(v_m) - J_m(v_0), v_m - v_0) \to 0.$$

Now by Lemma 2.2 we have

$$\int\limits_\Omega \left[I(v_m)^{1/2} - I(v_0)^{1/2} \right]^2 dx \to 0. \tag{3.26}$$

Since the function $u \mapsto \int\limits_\Omega u^2\, dx$ is a continuous mapping from $L_2(\Omega)$ into $\mathbb{R}$, we obtain (3.23) from (3.26). ∎

Remark. From the proof of Theorem 3.1 it can be seen that there exists a solution of problem (2.13) when the constant a_5 in (2.8) is equal to zero. But in this case subsequence $\{v_m\}$ converges to v weakly in V.

4. Nonstationary problem

4.1 Formulation of the problem

We take the constitutive equation of the form

$$\sigma_{ij}(p,v) = -p\,\delta_{ij} + [f(\|v\|^2) + 2\varphi(I(v),\|v\|_2^2)]\varepsilon_{ij}(v). \tag{4.1}$$

Note that equation (4.1) differs from (2.1) by the addend $f(\|v\|^2)$ in the viscosity. However, it may be that $f(z) = 0 \ \forall z \in [0,b_0]$ and $y_1, y_2 \to \varphi(y_1, y_2)$ is independent of y_2 for $y_2 \in [0,b_1]$, where b_0, b_1 are arbitrary positive constants (see (4.10), (4.14)). State equation (4.1) may be also used for stationary problems.

Let Ω be a bounded domain in $\mathbb{R}^n$ $(n = 2,3)$ with a Lipschitz continuous boundary S, and S_1, S_2 be open non-empty subsets in S such that $S = \bar{S}_1 \cup \bar{S}_2$, and $S_1 \cup S_2$ is an empty set.

The problem on stationary flow of a fluid with the constitutive equation (4.1) consists in finding a pair of functions (p,v) satisfying

$$\rho\frac{\partial v_i}{\partial t} + \rho v_j \frac{\partial v_i}{\partial x_j} - \frac{\partial \sigma_{ij}(p,v)}{\partial x_j} = K_i \qquad \text{in } \Omega \times (0,T), \qquad i = 1,...,n, \tag{4.2}$$

$$\operatorname{div} v = 0 \qquad \text{in } \Omega \times (0,T), \tag{4.3}$$

$$v\big|_{S_1 \times (0,T)} = 0, \tag{4.4}$$

$$\sigma_{ij}(p,v)\nu_j\big|_{S_2 \times (0,T)} = F_i \qquad i = 1,...,n, \tag{4.5}$$

$$v(x,0) = v_0(x) \qquad x \in \Omega. \tag{4.6}$$

Here T is an arbitrary positive constant.

Let now

$$U = \{u \in C^\infty(\Omega)^n, \quad u\big|_{S_1} = 0, \quad \operatorname{div} u = 0\}, \tag{4.7}$$

$$H \quad \text{be the closure of } U \text{ in } L_2(\Omega)^n, \tag{4.8}$$

We suppose that

$$\left.\begin{array}{l} \varphi \text{ is a continuous function on } \mathbb{R}_+^2 \text{ satisfying Lipschitz} \\ \text{condition in an arbitrary bounded subset of } \mathbb{R}_+^2 \end{array}\right\} \tag{4.9}$$

$$a_2 \geq \varphi(y_1, y_2) \geq a_1 \qquad \forall\, (y_1, y_2) \in \mathbb{R}_+ \times [0, a_0] \qquad (4.10)$$

$$a_3 y_2^k \geq \varphi(y_1, y_2) \geq a_1 \qquad \forall\, (y_1, y_2) \in \mathbb{R}_+ \times (a_0, \infty) \qquad (4.11)$$

$$[\varphi(z_1^2, y)z_1 - \varphi(z_2^2, y)z_2](z_1 - z_2) \geq 0 \qquad \forall\, (z_1, z_2, y) \in \mathbb{R}_+^3, \qquad (4.12)$$

where $a_0 - a_3$, k are arbitrary positive constants.

We take the following assumptions for f:

$$\left.\begin{array}{l} f \text{ is a continuous function on } \mathbb{R}_+ \text{ satisfying the Lipschitz} \\ \text{condition in an arbitrary bounded subset of } \mathbb{R}_+ \end{array}\right\} \qquad (4.13)$$

$$f(z) \geq 0 \quad \forall\, z \in \mathbb{R}_+, \qquad f(z_1) \geq f(z_2) \qquad \text{if } z_1 \geq z_2, \qquad (4.14)$$

$$a_4 z \leq f(z) \leq a_5 z \qquad \text{for } z \in [z_0, \infty], \qquad (4.15)$$

where a_4, a_5, z_0 are positive constants. We suppose also that

$$K, F \in L_2(0, T; V^*), \qquad (4.16)$$

$$v_0 \in H, \qquad (4.17)$$

the space V being defined by (2.10). Define operators N_1, N_2, N which map V into V^* as follows

$$(N_1(u), h) = f(\|u\|^2) \int_\Omega \varepsilon_{ij}(u)\varepsilon_{ij}(h)\, dx, \qquad (4.18)$$

$$(N_2(u), h) = 2 \int_\Omega \varphi(I(u), \|u\|_2^2)\varepsilon_{ij}(u)\varepsilon_{ij}(h)\, dx, \qquad (4.19)$$

$$N = N_1 + N_2 + L_2, \qquad (4.20)$$

where L_2 is defined by (2.30).

Consider the problem: *find a function v satisfying*

$$v \in L_4(0, T; V) \cap L_\infty(0, T; H), \qquad \frac{dv}{dt} \in L_{4/3}(0, T; V^*), \qquad (4.21)$$

$$\rho\left(\frac{dv}{dt}, h\right) + (N(v), h) = (K + F, h) \qquad \forall\, h \in V, \qquad (4.22)$$

$$v(0) = v_0. \qquad (4.23)$$

It is easy to see that if (p, v) is a smooth solution of problem (4.1)–(4.6) then v is a solution of problem (4.21)–(4.23). On the contrary, by analogy with [12, Theorem 1.3, p. 160] it may be shown that if v is a solution of problem (4.21)–(4.23), then there exists a function $p \in D'(\Omega \times (0, T))$ such that the pair (p, v) is a solution of problem (4.1)–(4.6) in the sense of distributions.

4.2 Approximate solutions and a priori estimates

Let $\{V_n\}$ be a sequence of finite dimensional subspaces in V satisfying (3.1), (3.2) and $\{w_i\}_{i=1}^{k(n)}$ be a basis in V_n. We also suppose that

$$V_n \subset W_\infty^1(\Omega)^n \qquad (4.24)$$

Consider the problem: *find a function* $v_n : t \to v_n(t)$ *of the form* $v_n(t) = \sum_{i=1}^{k(n)} g_{in}(t) w_i$ *that satisfies*

$$\rho \frac{d}{dt}(v_n(t), w_i) + (N(v_n(t)), w_i) = (K(t) + F(t), w_i) \qquad i = 1, \ldots, k(n) \quad (4.25)$$

$$v_n(0) = v_{0n} \in V_n \qquad v_{0n} \to v_0 \quad \text{in } H. \qquad (4.26)$$

4.1 Lemma. *Let the conditions (4.9)–(4.17) be satisfied. Then there exists a unique solution of problem (4.25), (4.26) and the following estimates hold:*

$$\|v_n\|_{L_4(0,T;V)} \le c \qquad \forall n, \qquad (4.27)$$

$$\|v_n\|_{L_\infty(0,T;H)} \le c_1 \qquad \forall n. \qquad (4.28)$$

PROOF :   The matrix $((w_i, w_j))$ is invertible because $\{w_i\}_{i=1}^{k(n)}$ is a basis in V_n, and so (4.25) is a system of the form

$$\frac{d}{dt} g_{in}(t) = \psi_{in}(t, g_{in}(t), \ldots, g_{k(n)}(t)) \qquad i = 1, \ldots k(n), \qquad (4.29)$$

with the conditions

$$g_{in}(0) = g_{in0} \qquad i = 1, \ldots, k(n), \qquad (4.30)$$

and $v_{0n} = \sum_{i=1}^{k(n)} g_{in0} w_i$. The functions ψ_{in} satisfy a Lipschitz condition in an arbitrary bounded vicinity of zero, and so there exists a unique local solution of problem (4.29), (4.30) on some subinterval $[0, t_n]$ of the segment $[0, T]$. If we prove the estimate (4.25), then we shall obtain $t_n = T$. We multiply equation (4.25) with the function g_{in} and summarize over i. This gives

$$\psi\big(\frac{d}{dt} v_n(t), v_n(t)\big) + (N(v_n(t)), v_n(t)) = (K(t) + F(t), v_n(t)). \qquad (4.31)$$

Next, let us integrate both sides of this equation over t from zero to t and apply Green formula. Taking into account (4.18)–(4.20) we obtain

$$\frac{1}{2}\rho |v_n(t)|_2^2 + \int_0^t \int_\Omega \left[f(|v_n(\tau)|^2) + 2\varphi(I(v_n(\tau)), |v_n(\tau)|_2^2) \right] II(v_n(t)) \, dx \, d\tau +$$

$$\qquad (4.32)$$

$$+ \rho \int_0^t \int_\Omega v_{ni}(\tau) \, dx \, d\tau = \frac{1}{2}\rho |v_n(0)|_2^2 + \int_0^t (K(\tau) + F(\tau), v_n(\tau)) \, d\tau.$$

We take the notation

$$A_n(t) = \int\limits_0^t \int_\Omega \left\{ \left[\frac{1}{2} f(|v_n(\tau)|^2) + \varphi(I(v_n(\tau)), |v_n(\tau)|_2^2) \right] I(v_n(\tau)) + \right.$$

$$\left. + \rho v_{nj}(\tau) \frac{\partial v_{ni}}{\partial x_j}(\tau) v_{ni}(\tau) \right\} \, dx \, d\tau. \tag{4.33}$$

Taking into account (4.10), (4.11), (4.14), (4.15) and applying the Hölder inequality we get

$$A_n(t) \geq \int\limits_0^t \left(e_n(\tau) |v_n(\tau)|^2 - c_1 |v_n(\tau)| \, |v_n(\tau)|_4^2 \right) d\tau, \tag{4.34}$$

where $c_1 > 0$ and

$$e_n(\tau) = \begin{cases} a_1 & \text{if } |v_n(\tau)|^2 \leq z_0 \\ \frac{1}{2} a_4 |v_n(\tau)|^2 & \text{if } |v_n(\tau)|^2 > z_0. \end{cases} \tag{4.35}$$

It follows from (4.34), (4.35) that for all n and for all $t \in [0, T]$

$$A_n(t) \geq -c_2, \qquad c_2 = \text{const} > 0. \tag{4.36}$$

We have

$$\left| \int_0^t (K(\tau) + F(\tau), v_n(\tau)) \, d\tau \right| \leq \int_0^t \|K(\tau) + F(\tau)\|_{V^*} \, \|v_n(\tau)\| \, d\tau \leq$$

$$\leq \frac{1}{2\varepsilon} \int_0^t \|K(\tau) + F(\tau)\|_{V^*}^2 \, d\tau + \frac{\varepsilon}{2} \int_0^t \|v_n(\tau)\|^2 \, d\tau \tag{4.37}$$

where ϵ is an arbitrary positive constant. We take $\epsilon = a_1$, then we obtain from (4.10), (4.11), (4.32), (4.36)

$$\frac{1}{2} \rho \|v_n(t)\|_2^2 + \frac{1}{2} \int\limits_0^t f(\|v_n(\tau)\|^2) \|v_n(\tau)\|^2 \, d\tau + \frac{1}{2} a_1 \int\limits_0^t \|v_n(\tau)\|^2 \, d\tau \leq$$

$$\leq \frac{1}{2} \rho \|v_n(0)\|_2^2 + \frac{1}{2a_1} \int\limits_0^t \|K(\tau) + F(\tau)\|_{V^*}^2 \, d\tau + c_2. \tag{4.38}$$

Now (4.28) follows from (4.38) due to (4.14), (4.16), (4.26). By (4.38), (4.14), (4.15) we obtain (4.27). ∎

4.2 Lemma. *Under conditions (4.13), (4.14) the operator N_1 defined by (4.18) is a continuous mapping from V into V^* satisfying*

$$(N_1(u) - N_1(w), u - w) \geq 0 \qquad \forall u, w \in V. \tag{4.39}$$

PROOF : It follows from (4.18), (4.14) that

$$(N_1(u) - N_1(w), u - w) =$$

$$= \int_\Omega \left[f(\|u\|^2)\epsilon_{ij}(u) - f(\|w\|^2)\epsilon_{ij}(w) \right] \epsilon_{ij}(u - w) \, dx = \tag{4.40}$$

$$= \int_\Omega \left[f(\|u\|^2)I(u) + f(\|w\|^2)I(w) - (f(\|u\|^2) + f(\|w\|^2))\epsilon_{ij}(u)\epsilon_{ij}(w) \right] dx \geq$$

$$\geq f(\|u\|^2)\|u\|^2 + f(\|w\|^2)\|w\|^2 - (f(\|u\|^2) + f(\|w\|^2))\|u\| \, \|w\| =$$

$$= (f(\|u\|^2)\|u\| - f(\|w\|^2)\|w\|)(\|u\| - \|w\|) \geq 0.$$

The continuity of the operator N_1 is obvious. ∎

4.3 Lemma. *Under conditions (4.19)–(4.17) the next estimate holds*

$$\|n_1(v_n)\|_{L_{4/3}(0,T;V^*)} \leq c \qquad \forall n. \tag{4.41}$$

PROOF : We introduce the function

$$g_n(t) = \begin{cases} f(z_0) & \text{if } \|v_n(t)\|^2 \leq z_0 \\ a_5\|v_n(t)\|^2 & \text{if } \|v_n(t)\|^2 > z_0. \end{cases} \tag{4.42}$$

By (4.13)–(4.15) we get

$$\left| \int_0^T (N_1(v_n(t)), h(t)) \, dt \right| = \left| \int_0^T f(\|v_n(t)\|^2) \int_\Omega \epsilon_{ij}(v_n(t))\epsilon_{ij}(h(t)) \, dx \, dt \right| \leq$$

$$\leq \int_0^T g_n(t)\|v_n(t)\| \, \|h(t)\| \, dt \leq \left(\int_0^T (g_n(t)\|v_n(t)\|)^{4/3} \, dt \right)^{3/4} \|h\|_{L_4(0,T;V)}. \tag{4.43}$$

From (4.27) and (4.42) it follows that

$$\int_0^T (g_n(t)\|v_n(t)\|)^{4/3} \, dt \leq c_1 \qquad \forall n, \tag{4.44}$$

and (4.43), (4.44) imply (4.41).

4.4 Lemma. *Let the conditions (4.9)–(4.17) be satisfied. Then*

$$\|N_2(v_n)\|_{L_2(0,T;v^*)} \leq c \qquad \forall n. \tag{4.45}$$

Indeed, by (4.19), (4.10), (4.11) and (4.28) we get

$$\Big| \int\limits_0^T (N_2(v_n(t)), h(t))\, dt \Big| =$$

$$= 2\Big| \int\limits_0^T \int_\Omega \varphi(I(v_n(t)), \|v_n(t)\|_2^2)\epsilon_{ij}(v_n(t)) \times \times \epsilon_{ij}(h(t))\, dx\, dt \Big| \leq$$

$$\leq c \int\limits_0^T \|v_n(t)\|\, \|h(t)\|\, dt \leq c\|v_n\|_{L_2(0,T;V)}\|h\|_{L_2(0,T,V)}.$$

From here and (4.27) we obtain (4.45).

4.5 Lemma. *Let the conditions (4.9)–(4.17) be satisfied, and the operator L_2 be defined by (2.30). Then*

$$\|L_2(v_n)\|_{L_2(0,T;V^*)} \leq c \qquad \forall n. \tag{4.46}$$

PROOF : By the Hölder inequality and the embedding theorem we get

$$\Big| \int_0^T \big(L_2(v_n(t)), h(t) \big)\, dt \Big| \leq c_1 \int_0^T \|v_n(t)\|_4 \|v_n(t)\| \|h(t)\|_4\, dt$$

$$\leq c_2 \int_0^T \|v_n(t)^2 \|h(t)\|\, dt \leq c_2 \|v_n\|_{L_4(0,T;V)}^2 \|h\|_{L_2(0,T;V)}.$$

From here and (4.27) we obtain (4.46). ■

4.6 Lemma. *Let the conditions (4.9)–(4.17) be satisfied. Then*

$$\Big\|\frac{dv_n}{dt}\Big\|_{L_{4/3}(0,T;V^*)} \leq c. \tag{4.47}$$

PROOF : Let h be an arbitrary element of V. This element is uniquely presented in the form $h = h_1 + h_2$, where $h_1 \in V_n$ and h_2 is orthogonal to V_n concerning the scalar product in H. Therefore, equations (4.26) are presented as follows

$$\rho\frac{dv_n}{dt}(t) = -P_n N(v_n(t)) + P_n(K(t) + F(t)), \tag{4.48}$$

where P_n is the projector in V^* onto V_n^*, $\|P_n\|_{\mathcal{L}(V^*, V_n^*)} = 1$. Now by (4.16), (4.41), (4.45), (4.46), (4.48) we obtain (4.47). ∎

Now let B_0, B, B_1 be Banach spaces such that

$$\left. \begin{array}{l} B_0 \subset B \subset B_1, B_0 \text{ and } B_1 \text{ are reflexive ,} \\ \text{the embedding } B_0 \to B \text{ is compact.} \end{array} \right\} \tag{4.49}$$

Let also

$$W = \left\{ u \in L_{p_0}(0, T; B_0), \frac{du}{dt} \in L_{p_1}(0, T; B_1) \right\}$$

where $1 < p_i < \infty$ $i = 0, 1$ and t is finite. The set W supplied with the norm

$$\|u\|_W = \|u\|_{L_{p_0}(0, T; B_0)} + \left\| \frac{du}{dt} \right\|_{L_{p_1}(0, T; B_1)}$$

is a Banach space.

We shall use the following known result [3].

4.7 Lemma. *Under condition (4.49) the embedding $W \mapsto L_{p_0}(0, T; B)$ is compact.*

5. Existence theorem

We prove the following result.

5.1 Theorem. *let the conditions (4.9)–(4.17) hold. Then there exists a solution v of the problem (4.21)–(4.23), and from the sequence $\{v_n\}$ of solutions of problem (4.25), (4.26) a subsequence $\{v_l\}$ may be extracted such that $v_l \to v$ weakly in $L_4(0, T; V)$ and $v_l \to v$ in $L_4(Q)$, where $Q = \Omega \times (0, T)$.*

PROOF : **1.** It follows from Lemmata 4.1, 4.3, 4.4 that there exists a subsequence $\{v_l\}$ of $\{v_n\}$ such that

$$\begin{array}{lll} v_l \to v & \text{weakly in } L_4(0, T; V), & (5.1) \\ v_l \to v & \text{weakly}^* \text{ in } L_\infty(0, T; H), & (5.2) \\ v_l(T) \to y & \text{weakly in } H, & (5.3) \\ N_1(v_l) \to Z_1 & \text{weakly in } L_{4/3}(0, T; V^*), & (5.4) \\ N_2(v_l) \to Z_2 & \text{weakly in } L_2(0, T; V^*). & (5.5) \end{array}$$

Due to (4.27), (4.47) and Lemma 4.7 we may consider that

$$\left. \begin{array}{l} v_l \to v \text{ strongly in } L_4(0, T; L_4(\Omega)^n) = L_4(Q)^n \\ \text{and almost everywhere in } Q. \end{array} \right\} \tag{5.6}$$

Multiplying equation (4.25) for $n = l$ with a function $\nu \in C^1([0, T])$, integrating the result from 0 to T and using the Green formula, we obtain

$$
\int_0^T \left[-\rho(v_l(t), w_i) \frac{d\nu}{dt}(t) + (N_1(v_l(t)), w_i)\nu(t) + (N_2(v_l(t)), w_i)\nu(t) + \right.
$$
$$
\left. + (L_2(v_l(t)), w_i)\nu(t) \right] dt = \int_0^T \left(K(t) + F(t), w_i \right)\nu(t)\, dt \tag{5.7}
$$
$$
+ \rho(v_l(0), w_i)\nu(0) - \rho(v_l(T), w_i)\nu(T) \qquad i = 1, \ldots, k(l).
$$

Since $w_i \in L_\infty(\Omega)^n$ (see (4.24)) it follows from (5.1) and (5.6) that

$$
\lim_{l \to \infty} \int_0^T \left(L_2(v_l(t)), w_i \right)\nu(t)\, dt = \int_0^T \left(L_2(v(t)), w_i \right)\nu(t)\, dt. \tag{5.8}
$$

Now for fixed w_i we pass to the limit in (5.7). Using (4.28), (5.1), (5.3)–(5.5), (5.8) we get

$$
\int_0^T \left[-\rho(v(t), w_i) \frac{d\nu}{dt}(t) + (Z_1(t) + Z_2(t), w_i)\nu(t) + +(L_2(v(t)), w_i)\nu(t) \right] dt =
$$
$$
\int_0^T \left(K(t) + F(t), w_i \right)\nu(t)\, dt + \rho(v_0, w_i)\nu(0) - \rho(y, w_i)\nu(T). \tag{5.9}
$$

Since $D(0, T) \subset C^1([0, T])$ it follows from (3.1) and (5.9) that

$$
\rho \frac{dv}{dt} + Z_1 + Z_2 + L_2(v) = K + F. \tag{5.10}
$$

From Lemma 4.5 and 5.1 it follows that $L_2(v) \in L_2(0, T; V^*)$, and by (5.10), (4.16), (5.4), (5.5) we get $\dfrac{dv}{dt} \in L_{4/3}(0, T; V^*)$. Therefore we may integrate (5.9) by parts. Taking a function $\nu \in C^1([0, T])$, $\nu(0) = 0$ from (5.9), (5.10) we obtain

$$
v(T) = y, \tag{5.11}
$$

and taking ν such that $\nu(T) = 0$, we have

$$
v(0) = v_0. \tag{5.12}
$$

98

2. We shall show now that

$$Z_1 + Z_2 = N_1(v) + N_2(v). \tag{5.13}$$

Use the notation

$$X_l(u) = \int_0^T \int_\Omega \left\{ [f(\|v_l\|^2) + 2\varphi(I(v_l), \|v_l\|_2^2)]\varepsilon_{ij}(v_l) - \right.$$
$$\left. - [f(\|u\|^2) + 2\varphi(I(u), \|v_l\|_2^2)]\varepsilon_{ij}(u) \right\} \varepsilon_{ij}(v_l - u) \, dx \, dt, \tag{5.14}$$

where $u \in L_4(0, T; V)$. By Lemmata 2.2 and 4.2 we have

$$X_l(u) \geq 0 \qquad \forall u \in L_4(0, T; V), \forall l. \tag{5.15}$$

We take $t = T, n = l$ into (4.32), this gives

$$\int_0^T \int_\Omega [f(\|v_l\|^2) + 2\varphi(I(v_l), \|v_l\|_2^2)]I(v_l) \, dx \, dt =$$
$$= \int_0^T (K + F, v_l) \, dt + \frac{1}{2}\rho\|v_l(0)\|_2^2 - \frac{1}{2}\rho\|v_l(T)\|_2^2 - \int_0^T (L_2(v_l), v_l) \, dt. \tag{5.16}$$

By (5.14) and (5.16) we get

$$X_l(u) = \frac{1}{2}\rho\|v_l(0)\|_2^2 - \frac{1}{2}\rho\|v_l(T)\|_2^2 + \int_0^T (K + F, v_l) \, dt - \int_0^T (L_2(v_l), v_l) \, dt -$$
$$- \int_0^T \int_\Omega [f(\|v_l\|^2) + 2\varphi(I(v_l), \|v_l\|_2^2)]\varepsilon_{ij}(v_l)\varepsilon_{ij}(u) \, dx \, dt -$$
$$- \int_0^T \int_\Omega \{[f(\|u\|^2) + 2\varphi(I(u), \|v_l\|_2^2)]\varepsilon_{ij}(v_l)\varepsilon_{ij}(v_l - u) \, dx \, dt. \tag{5.17}$$

Due to (5.3), (5.11) we have

$$\lim_{l \to \infty} \|v_l(T)\|_2 \geq \|v(T)\|_2. \tag{5.18}$$

Taking (4.10), (4.11), (5.2), (5.6) into account and applying the Lebesgue theorem we obtain

$$\varphi(I(u), \|v_l\|_2^2)\varepsilon_{ij}(u) \to \varphi(I(u), \|v\|_2^2)\varepsilon_{ij}(u) \qquad \text{in } L_2(Q).$$

This relation together with (5.1) gives

$$\lim_{l \to \infty} \int_0^T \int_\Omega \varphi(I(u), \|v_l\|_2^2) \varepsilon_{ij}(u) \varepsilon_{ij}(v_l - u) \, dx \, dt =$$

$$= \int_0^T \int_\Omega \varphi(I(u), \|v\|_2^2) \varepsilon_{ij}(u) \varepsilon_{ij}(v - u) \, dx \, dt. \tag{5.19}$$

We have

$$\|v_{li}v_{lj} - v_i v_j\|_{L_2(Q)} \leq \|(v_{li} - v_i)v_{lj}\|_{L_2(Q)} + \|v_i(v_{lj} - v_j)\|_{L_2(Q)}, \tag{5.20}$$

and next, by Hölder inequality, we obtain

$$\|(v_{li} - v_i)v_{lj}\|_{L_2(Q)} \leq \|(v_{li} - v_i)\|_{L_4(Q)} \|v_{li}\|_{L_4(Q)}. \tag{5.21}$$

Now (5.6), (5.20), (5.21) give $v_{li}v_{lj} \to v_i v_j$ in $L_2(Q)$, and so by (5.1) we obtain

$$\lim_{l \to \infty} \int_0^T \left(L_2(v_l), v_l \right) dt = \int_0^T \left(L_2(v), v \right) dt. \tag{5.22}$$

From (5.15), (5.17) applying (4.26), (5.1), (5.4), (5.5), (5.12), (5.18), (5.19), (5.22) we get

$$0 \leq \overline{\lim_{l \to \infty}} X_l(u) + \frac{1}{2}\rho\|v(0)\|_2^2 - \frac{1}{2}\rho\|v(T)\|_2^2 \leq$$

$$\leq \int_0^T (K + F, v) \, dt - \int_0^T \left(L_2(v), v \right) dt - \int_0^T (Z_1 + Z_2, u) \, dt - \tag{5.23}$$

$$- \int_0^T \int_\Omega \left[f(\|u\|^2) + 2\varphi(I(u), \|v\|_2^2) \right] \varepsilon_{ij}(u) \varepsilon_{ij}(v - u) \, dx \, dt$$

for all $u \in L_4(0, T; V)$. Multiplying (5.10) with v in the sense of the duality between V and V^*, and integrating with respect to t from 0 to T we obtain

$$\int_0^T (K + F, v) \, dt + \frac{1}{2}\rho\|v(0)\|_2^2 - \frac{1}{2}\rho\|v(T)\|_2^2 - \int_0^T \left(L_2(v), v \right) dt = \int_0^T (Z_1 + Z_2, v) \, dt.$$

From here and (5.23) we get for all $u \in L_4(0, T; V)$

$$- \int_0^T (Z_1 + Z_2, v - u) \, dt - \int_0^T \int_\Omega \left[f(\|u\|^2) + 2\varphi(I(u), \|v\|_2^2) \right] \varepsilon_{ij}(u) \varepsilon_{ij}(v - u) \, dx \, dt \geq 0.$$

We take here $u = v - \lambda w$, $\lambda > 0$ and w is an arbitrary element of $C([0,T];V)$. By applying Lemma 4.2 and the Lebesgue theorem we pass to the limit as $\lambda \to 0$. This yields

$$- \int\limits_0^T (Z_1 + Z_2, w)\, dt - \int\limits_0^T \int\limits_\Omega \left[f(\|v\|^2) + 2\varphi(I(v), \|v\|_2^2) \right] \varepsilon_{ij}(v)\, \varepsilon_{ij}(w)\, dx\, dt \geq 0 \quad (5.24)$$

for all $w \in C([0,T];V)$. Because $C([0,T];V)$ is dense in $L^2(0,T;V)$, we obtain (5.13) by (4.18), (4.19), (5.24). ∎

Acknowledgement

Part of this work was supported by the fund for the fundamental research of Ukrainian State Committee on Science and Technology, grant No. 1/364.

References

[1] W.G. Litvinov, *The motion of nonlinearly-viscous fluid*, Nauka, Moscow, 1982.

[2] O.A. Ladyzhenskaya, *The mathematical theory of viscous incompressible flow*, Gordon and Breach, New York, 1969.

[3] J.L. Lions, *Quelques méthodes de résolution des problèmes aux limites non linéaires*, Dunod, Paris, 1969.

[4] R. Temam, *Navier–Stokes equations*, North–Holland, Amsterdam, 1977.

[5] N.D. Kopachevsky, S.G. Krein, *On one problem of flow of viscous fluid*, ZAA **8(6)** (1989), 557–561.

[6] A.G. Fredrickson, *Principles and applications of rheology*, Prentice–Hall, Inc. Englewood Cliffs, 1964.

[7] S. Middleman, *The flow of high polymers*, Interscience Publishing, New York, 1968.

[8] G. Astarita, G. Marrucci, *Principles of non–Newtonian fluid mechanics*, McGraw–Hill, New York, 1974.

[9] L.G. Loitzansky, *Mechanics of the fluid and gas*, Nauka, Moscow, 1987.

[10] O.A. Ladyzhenskaya, *On modifications of the Navier–Stokes equations for high velocity gradients*, LOMI **7** (1968), 126–154.

[11] M.I. Vishik, A.V. Fursikov, *Mathematical problems of statistic hydromechanics*, Nauka, Moscow, 1980.

[12] V. Girault, P.A. Raviart, *Finite element approximation of the Navier–Stokes Equations*, Springer, Berlin, 1981.

[13] O.A. Ladyzhenskaya, V.A. Solonnikov, *Some problems of vector analysis and generalized formulation of boundary value problems for Navier–Stokes equations*, LOMI **59** (1976), 81–116.

[14] I. Mogilevskij, V.A. Solonnikov, *On the solvability of an evolution free boundary problem for the Navier–Stokes equations in Hölder spaces of functions*, Mathematical problems relating to the Navier–Stokes equations, (G.P. Galdi ed.) Series of Advances in Applied Math., vol. II, World Scientific, 1992.

William G. Litvinov
Institute of Mechanics
Academy of Science of Ukraine
Nesterov St. 3
252057 Kiev
Ukraine

AKITAKA MATSUMURA

Large-Time Behavior of One–Dimensional Motions of Compressible Viscous Gas

Abstract

Through this series of lectures, we survey some recent developments on the global existence and large-time behavior of the solutions to the Cauchy problem for a one-dimensional model system of compressible viscous gas. First, taking a simple scalar equation, we recall an elementary but technical energy method to provide the global existence and large-time behavior toward the viscous shock waves and rarefaction waves. Next, as applications of this energy method, we study a barotropic model system in connection with the Riemann problem for the corresponding inviscid system. In particular, we show the stability of viscous shock waves without genuine nonlinearity, and the global asymptotics toward the rarefaction waves. Finally, we present some related topics and further problems from the initial boundary value problems.

1. Energy Method

In this section, we study a scalar viscous conservation law

$$u_t + f(u)_x = \mu u_{xx},\qquad(1.1)$$

where μ is a positive constant. We assume that f is a smooth convex function, i.e.,

$$f''(u) > 0 \quad \text{for } u \in \mathbb{R}.\qquad(1.2)$$

A typical example of (1.1) is the Burgers equation

$$u_t + uu_x = \mu u_{xx}.\qquad(1.3)$$

We consider the Cauchy problem for (1.1) with the initial data

$$u(0, x) = u_0(x) \quad x \in \mathbb{R},$$
$$\lim_{x \to \pm\infty} u_0(x) = u_\pm,\qquad(1.4)$$

where $u_\pm$ are given constants ($u_- \neq u_+$). The various large-time behavior for the solutions of (1.1), (1.4) were already well studied in Il'in and Oleinik [8]: roughly speaking, if $u_- > u_+$ then the solution tends toward a viscous shock wave solution of

(1.1), and if $u_- < u_+$, then it tends toward a rarefaction wave of the corresponding inviscid equation. However, since all arguments in [8] are due to the maximum principle, we cannot straightforwardly apply them to the system case. The arguments in the system case have been developed rather recently through quite a different approach which is based upon an elementary but technical energy method. This approach to the stability of the viscous shock waves was first introduced independently by Matsumura–Nishihara [27] and Goodman [5], and in the case of the rarefaction waves by Matsumura and Nishihara [28]. After these papers, there have been many works on the interesting systems both in the theory of gas dynamics and in more general frameworks ([12], [13], [19], [20], [21], [22], [23], [29], [39], [43]). In order to expose these energy methods clearly, we apply them to a relatively simple problem (1.1), (1.4).

1.1. Stability of Viscous Shock Waves

Let $u_\pm$ and s be constants. If $u = U(x - st)$ is a smooth solution of (1.1) satisfying $U(\pm\infty) = u_\pm$, then we call $U(x - st)$ a viscous shock wave which connects u_- and u_+. We can easily check the existence of a viscous shock wave $U(x - st)$ connecting u_- and u_+ if and only if $u_\pm$ and s satisfy the Rankine-Hugoniot condition

$$-s(u_+ - u_-) + (f(u_+) - f(u_-)) = 0 \tag{1.5}$$

and the shock condition in the sense of Lax [18]

$$f'(u_+) < s < f'(u_-), \tag{1.6}$$

which is equivalent to $u_- > u_+$ under the assumption (1.2). We can also see that $U(\xi)$ ($\xi = x - st$) is unique up to a shift ξ and a monotone function of ξ, that is,

$$U'(\xi) < 0 \quad \text{for} \quad \xi \in \mathbb{R}. \tag{1.7}$$

Let $U(x - st)$ is a viscous shock wave connecting u_- and u_+ and let us define α by

$$\int u_0(x) - U(x)dx = \alpha(u_+ - u_-), \tag{1.8}$$

where α is uniquely determined provided that $u_0 - U$ is integrable over $\mathbb{R}$. It is noted that the shifted $U(x - st + \alpha)$ is also a viscous shock wave connecting u_- and u_+ , and satisfies

$$\int u_0(x) - U(x + \alpha)dx = 0. \tag{1.9}$$

This viscous shock wave $U(x - st + \alpha)$ is proved to be stable as $t \to \infty$, provided the following φ_0 is suitably small in H^2 :

$$\varphi_0(x) = \int_{-\infty}^{x} (u_0(y) - U(y + \alpha))dy \in H^2. \tag{1.10}$$

Here H^k ($k \geq 0$) denotes the usual Sobolev spaces with the norm $\| \cdot \|_k$. We note $L^2 = H^0$ and simply set $\| \cdot \| = \| \cdot \|_0$.

Theorem 1.1. *Suppose $u_\pm$ and s satisfy (1.5), (1.6) and $u_0 - U$ is integrable over $\mathbb{R}$. We define α by (1.8) and assume (1.10). Then there is a constant $\varepsilon_0 > 0$ such that if $\|\varphi_0\|_2 \leq \varepsilon_0$ then the problem (1.1), (1.4) has a unique global solution u satisfying*

$$u - U \in C^0([0, \infty); H^1) \cap L^2(0, \infty; H^2) \tag{1.11}$$

where $U = U(x - st + \alpha)$, and the solution verifies

$$\sup_{x \in \mathbb{R}} |u(t, x) - U(x - st + \alpha)| \to 0 \quad \text{as } t \to \infty. \tag{1.12}$$

To prove Theorem 1.1, as in the previous papers [5], [27], we seek the solution of (1.1),(1.4) in the form

$$u(t, x) = U(x - st + \alpha) + \varphi_x(t, x). \tag{1.13}$$

Here φ is to be determined as a solution to the "integrated" problem :

$$\varphi_t + (f(U + \varphi_x) - f(U)) = \mu \varphi_{xx}, \tag{1.14}$$

$$\varphi(0, x) = \varphi_0(x) \in H^2. \tag{1.15}$$

We can prove the global solvability of this integrated problem by a combination of the following local existence and *a priori* estimate.

Proposition 1.2 (local existence). *For any $\tau \geq 0$ and $\varphi_\tau \in H^2$, consider the Cauchy problem for (1.14) with the initial data*

$$\varphi(\tau, x) = \varphi_\tau(x) \in H^2. \tag{1.16}$$

Then for any $M > 0$, there exists a $t_0 = t_0(M) > 0$ (independent of τ) such that if $\|\varphi_\tau\|_2 \leq M$, then the problem (1.14), (1.16) has a unique solution φ on $I = [\tau, \tau + t_0]$ satisfying

$$\varphi \in C^0(I; H^2), \varphi_x \in L^2(I; H^2),$$
$$\sup_{t \in I} \|\varphi(t)\|_2 \leq 2M. \tag{1.17}$$

Proposition 1.3 (*a priori* estimate). *We assume the same conditions as in the Theorem 1.1. Let φ be a solution of (1.14), (1.15) satisfying*

$$\varphi \in C^0([0, T]; H^2), \quad \varphi_x \in L^2(0, T; H^2) \tag{1.18}$$

for some $T > 0$, and put $N(t) = \sup_{0 \le s \le t} \|\varphi(s)\|_2$. Then there exist positive constants ε_1 and C_1 (both are independent of T) such that if $N(T) \le \varepsilon_1$, then it holds

$$\|\varphi(t)\|_2^2 + \int_0^t \|\varphi_x(s)\|_2^2 ds \le C_1 \|\varphi_0\|_2^2, \quad t \in [0, T]. \tag{1.19}$$

PROOF of Theorem 1.1: We define $\varepsilon_0 = \min(\varepsilon_1/2, \; \varepsilon_1/2\sqrt{C_1})$, $M_0 = \varepsilon_1/2$ and suppose $\|\varphi_0\|_2 \le \varepsilon_0$. Then Proposition 1.2 with $\tau = 0$, $\varphi_\tau = \varphi_0$ and $M = M_0$ ensures a unique solution φ on $[0, t_0(M_0)]$ satisfying $N(t_0) \le 2M_0 = \varepsilon_1$. So, Proposition 1.3 with $T = t_0$ implies $N(t_0) \le \sqrt{C_1} \|\varphi_0\|_2 \le \varepsilon_1/2$. Hence Proposition 1.1 with $\tau = t_0$, $\varphi_\tau = \varphi(t_0)$ and $M = M_0$ again gives a unique solution on $[t_0, 2t_0]$, eventually $[0, 2t_0]$, satisfying $N(2t_0) \le 2M_0 = \varepsilon_1$. So Proposition 1.3 with $T = 2t_0$ again implies $N(2t_0) \le \varepsilon_1/2$. Thus, repeating this continuation process, we can obtain a unique global solution φ satisfying (1.18) and the estimate (1.19) with $T = \infty$. We now define u by the formula (1.13). Then u becomes the desired global solution of (1.1), (1.4) satisfying (1.11). Finally, by virtue of the uniform estimate (1.19) and the equation (1.1), we can see both $\|\varphi_x(t)\|^2$ and its derivative with respect to t are integrable over $t \ge 0$. This implies $\|\varphi_x(t)\| \to 0$ as $t \to \infty$. With the aid of the Sobolev inequality, we have

$$\sup_{x \in \mathbb{R}} |u - U|^2 \le \sup_{x \in \mathbb{R}} |\varphi_x(t, x)|^2 \le 2\|\varphi_x(t)\| \|\varphi_{xx}(t)\| \to 0 \tag{1.20}$$

as $t \to \infty$. Thus the proof of Theorem 1.1 is completed. ∎

Since the proof of the local existence is standard, we only show how to derive the *a priori* estimate. We rewrite (1.1) in the form

$$\varphi_t + f'(U)\varphi_x - \mu \varphi_{xx} = F, \tag{1.21}$$

where $F = -(f(U + \varphi_x) - f(U) - f'(U)\varphi_x) = O(|\varphi_x|^2)$. Multiplying (1.21) by 2φ and integrating it over $[0, t] \times \mathbb{R}$, we have for $N(T) \le 1$ and $t \in [0, T]$

$$\|\varphi(t)\|^2 + 2\mu \int_0^t \|\varphi_x(s)\|^2 ds + \int_0^t \int -f''(U)U_x \varphi^2 dx ds$$
$$= \|\varphi_0\|^2 + \int_0^t \int 2\varphi F dx ds \le \|\varphi_0\|^2 + CN(t) \int_0^t \|\varphi_x(s)\|^2 ds, \tag{1.22}$$

where and in what follows C represents a generic positive constant which is independent of T. Therefore, by the monotone property of U (1.7) and the convexity of f (1.2), we obtain

$$\|\varphi(t)\|^2 + \int_0^t \|\varphi_x(s)\|^2 ds \le C\|\varphi_0\|^2, \tag{1.23}$$

for suitably small $N(T)$, say $N(T) \le \varepsilon_1$. Since the estimates for the higher derivatives are not essential but tedious, we omit the details.

Remark 1.4. Of course, in Il'in and Oleinik [8], there are no smallness assumptions for the initial data. They also showed that if $|\varphi_0(x)| \leq Ce^{-\alpha|x|}(\exists \alpha > 0)$, it holds $\sup_{x \in \mathbb{R}} |u - U| \leq Ce^{-\beta t}(\exists \beta > 0)$. In the particular case $f = u^2/2$ (Burgers equation (1.2)), Nishihara [35] showed further properties : if $|\varphi_0(x)| \leq C(1+|x|)^{-\alpha}$ ($\exists \alpha > 0$), it holds $\sup_{x \in \mathbb{R}} |u - U| \leq C(1+t)^{-\alpha}$ and this decay rate is optimal. These detailed results are not known for general f with convexity. However, for such f, Kawashima and Matsumura [12] showed that if $(1+|x|)^{\alpha/2}\varphi_0 \in H^2$ ($\exists \alpha > 0$) is suitably small, it holds $\sup_{x \in \mathbb{R}} |u - U| \leq C(1+t)^{-[\alpha]/2}$.

1.2. Asymptotics Toward Rarefaction Waves

In the case $u_- < u_+$, the solution of (1.1), (1.4) is expected to tend toward the solution of the corresponding Riemann problem :

$$u_t + f(u)_x = 0, \tag{1.24}$$

$$u(0, x) = u_0^R(x) \equiv \begin{cases} u_- & x < 0, \\ u_+ & x > 0. \end{cases} \tag{1.25}$$

In particular, the solution for the inviscid Burgers equation

$$\sigma_t + (\sigma^2/2)_x = 0, \tag{1.26}$$

$$\sigma(0, x) = \sigma_0^R(x) \equiv \begin{cases} \sigma_- & x < 0, \\ \sigma_+ & x > 0, \quad (\sigma_- < \sigma_+) \end{cases} \tag{1.27}$$

is given by $\sigma = \sigma^R(x/t; \sigma_-, \sigma_+)$ where

$$\sigma^R(\xi; \sigma_-, \sigma_+) = \begin{cases} \sigma_- & \xi < \sigma_-, \\ \xi & \sigma_+ < \xi < \sigma_-, \\ \sigma_+ & \xi < \sigma_-. \end{cases} \tag{1.28}$$

Then it is noted that the solution $u = u^R(x/t)$ of (1.24),(1.25) is determined by the formula $f'(u^R(\xi)) = \sigma^R(\xi; f'(u_-), f'(u_+))$.

Theorem 1.5. *Suppose $u_- < u_+$, $u_0 - u_0^R \in L^2$ and $u_{0,x} \in L^2$. Then there exists a positive constant ε_0 such that if $\|u_0 - u_0^R\| + \|u_{0,x}\| + |u_+ - u_-| \leq \varepsilon_0$, then the problem (1.1), (1.4) has a unique global solution $u(t, x)$ satisfying*

$$u - u^R \in C^0([0, \infty); L^2), u_x \in C^0([0, \infty); L^2) \cap L^2(0, \infty; H^1), \tag{1.29}$$

and the solution verifies

$$\sup_{x \in \mathbb{R}} |u(t, x) - u^R(x/t)| \to \infty \quad \text{as} \quad t \to \infty. \tag{1.30}$$

To prove Theorem 1.5, according to the arguments in [28], we first approximate $\sigma^R(x/t; \sigma_-, \sigma_+)$ by a smooth solution $\sigma(t, x; \sigma_-, \sigma_+)$ of the problem

$$\sigma_t + (\sigma^2/2)_x = 0, \tag{1.31}$$

$$\sigma(0, x) = \frac{1}{2}(\sigma_- + \sigma_+) + \frac{1}{2}(\sigma_+ - \sigma_-) \tanh(x). \tag{1.32}$$

By the standard method of characteristic curves, we have

Lemma 1.6. *Set $\delta = \sigma_+ - \sigma_- > 0$. There exists a unique smooth solution σ of (1.31), (1.32) satisfying*
  (i) $\sigma_- < \sigma(t, x) < \sigma_+$ and $\sigma_x(t, x) > 0$,
  (ii) $\forall p \geq 1 \exists C_p > 0$ s.t. $\|\sigma_x(t)\|_{L^p} \leq C_p \min(\delta, \delta^{1/p} t^{-1+1/p})$,
  (iii) $\forall p \geq 1 \exists C_p > 0$ s.t. $\|\sigma_{xx}(t)\|_{L^p} \leq C_p \min(\delta, t^{-1})$,
  (iv) $\forall t \geq 0, \sigma(t, x)$ rapidly tends toward $\sigma_\pm$ as $x \to \pm\infty$,
  (v) $\sup_{x \in \mathbb{R}} |\sigma(t, x) - \sigma^R(x/t)| \to 0$ as $t \to \infty$.

Now we make an approximate solution $U(t, x)$ to $u^R(x/t)$ by the formula $f'(U(t, x)) = \sigma(t, x; f'(u_-), f'(u_+))$. Then we can see that U is a smooth solution of (1.24) and satisfies all corresponding results in Lemma 1.6. If we set $u = U + \varphi$, then φ must satisfy the problem :

$$\varphi_t + (f(U + \varphi) - f(U))_x = \mu\varphi_{xx} + \mu U_{xx}, \tag{1.33}$$

$$\varphi(0, x) = \varphi_0(x) \equiv u_0(x) - U(0, x) \in H^1. \tag{1.34}$$

To prove the global solvability of the problem (1.33), (1.34), we make use of the combination of local existence and *a priori* estimate as in Section 1.1.

Proposition 1.7 (*a priori* estimate). *We assume the same conditions as in Theorem 1.5. Let φ be a solution of (1.33),(1.34) satisfying*

$$\varphi \in C^0([0, T]; H^1), \quad \varphi_x \in L^2(0, T; H^1), \tag{1.35}$$

for some $T > 0$, and put $N(t) = \sup_{0 \leq s \leq t} \|\varphi(s)\|_1$. Then there exist positive constants ε_1 and C_1 (both are independent of T) such that if $N(T) + |u_+ - u_-| \leq \varepsilon_1$, then it holds

$$\|\varphi(t)\|_1^2 + \int_0^t \|\varphi_x(s)\|_1^2 ds \leq C_1(\|\varphi_0\|_1^2 + |u_+ - u_-|^{1/6}), \quad t \in [0, T]. \tag{1.36}$$

We only show the first step of the proof of Proposition 1.7. We rewrite (1.33) in the form
$$\varphi_t + (f(U)\varphi)_x - \mu\varphi_{xx} = \mu U_{xx} - F,$$
$$F = (f(U + \varphi) - f(U) - f'(U)\varphi)_x. \tag{1.37}$$

Multiplying (1.37) by 2φ and integrating it over $[0, t] \times R$, it holds

$$\|\varphi(t)\|^2 + 2\mu \int_0^t \|\varphi_x(s)\|^2 \, ds + \int_0^t \int f''(U) U_x \varphi^2 dx ds =$$
$$= \|\varphi_0\|^2 + \int_0^t \int 2\mu \varphi U_{xx} + 2\varphi F dx ds \,. \tag{1.38}$$

By Lemma 1.6,

$$\left| \int \varphi U_{xx} dx \right| \leq \sup |\varphi| \|U_{xx}\|_{L^1} \leq C(\|\varphi\|^2 \|\varphi_x\|^2 + \|U_{xx}\|_{L^1}^{4/3})$$
$$\leq CN(T) \|\varphi_x\|^2 + C|u_+ - u_-|^{1/6}(1+t)^{-7/6} \,, \tag{1.39}$$

and

$$\left| \int \varphi F dx \right| \leq C \int |\varphi|^3 U_x dx \leq CN(T) \int U_x |\varphi|^2 dx \,, \tag{1.40}$$

for $N(T) + |u_+ - u_-| \leq 1$. Therefore, by the monotone property $U_x > 0$ and the convexity of f (1.2), we obtain

$$\|\varphi(t)\|^2 + \int_0^t \|\varphi_x(s)\|^2 ds \leq C(\|\varphi_0\|^2 + |u_+ - u|^{1/6}) \tag{1.41}$$

for suitably small $N(T)$ and $|u_+ - u_-|$. Along the same line as in Section 1.1, we can have the global solution φ of (1.33), (1.34) satisfying (1.35) and (1.36) with $T = \infty$, and furthermore $\|\varphi_x(t)\| \to 0$ as $t \to \infty$. So, we can finally prove the asymptotic behavior (1.30) by

$$\sup |u - u^R| \leq \sup |u - U| + \sup |U - u^R|$$
$$\leq \sqrt{2} \|\varphi\|^{1/2} \|\varphi_x\|^{1/2} + \sup |U - u^R| \to 0, \text{ as } t \to \infty \,. \tag{1.42}$$

Thus we can prove Theorem 1.5.

Remark 1.8. (i) In [8], there are no smallness assumptions for $|u_+ - u_-|$ and the initial data. As to the decay rate, Hattori and Nishihara [6] showed in the case of Burger's equation ($f = u^2/2$) the solution φ decays as $t^{-1/2}$ in a neighbourhood of the edges of the rarefaction wave $u^R(x/t)$, as t^{-1} otherwise, and furthermore these decay rates are optimal. This sort of detail decay estimates for general f with convexity is not known.

(ii) Matsumura-Nishihara [29] pointed out that if $\sigma(0, x)$ is replaced by

$$\sigma(0, x) = \tfrac{1}{2} (\sigma_- + \sigma_+) + \tfrac{1}{2} (\sigma_+ - \sigma_-) \kappa_p \int_0^{\varepsilon x} (1 + y^2)^{-q} dy$$
$$(\varepsilon > 0, q > 3/2, \kappa_p = \left(\int_0^\infty (1 + y^2)^{-q} \, dy \right)^{-1}) \tag{1.43}$$

instead of (1.32), it holds

$$\int_0^\infty \|\sigma_{xx}(t)\| dt \leq C|u_+ - u_-|^{-1/4q} < \infty, \tag{1.44}$$

and eventually the same estimate holds also for U_{xx}. By this property, in the case when $f = u^2/2$, we can show the large-time behavior (1.30) without any smallness assumption only by our energy method. In fact, formula (1.38) becomes in this case

$$\|\varphi(t)\|^2 + 2\mu \int_0^t \|\varphi_x(s)\|^2 ds + \int_0^t \int U_x \varphi^2 dx ds$$
$$= \|\varphi_0\|^2 + \int_0^t \int 2\mu \varphi U_{xx} dx ds \tag{1.45}$$
$$\leq \|\varphi_0\|^2 + 2\mu(\sup_{0 \leq s \leq t} \|\varphi(s)\|) C, \quad t \in [0, T]$$

which easily implies

$$\sup_{0 \leq s \leq t} \|\varphi(s)\|^2 + \int_0^t \|\varphi_x(s)\|^2 ds \leq C\|\varphi_0\|^2 + C(u_-, u_+) < \infty, \tag{1.46}$$

without any smallness conditions. This sort of technique shall play an essential role in the following Section 2.2.
(iii) Matsumura-Nishihara [30] also showed the asymptotic behavior (1.30) without any smallness conditions for the case where the viscosity term is nonlinearly degenerate (of a non-Newtonian type $\mu(|u_x|^{p-1}u_x)_x (p > 1)$).

2. Barotropic Model of Compressible Viscous Gas

In this section, we consider the large-time behavior of the solutions to the Cauchy problem for a barotropic model of compressible viscous gas :

$$v_t - u_t = 0,$$
$$u_t + p(v)_x = \mu(\frac{u_x}{v})_x, \tag{2.1}$$

with

$$(v, u)(0, x) = (v_0, u_0)(x),$$
$$\inf_{x \in \mathbb{R}} v_0(x) > 0, \quad \lim_{x \to \pm\infty} (v_0, u_0)(x) = (v_\pm, u_\pm), \tag{2.2}$$

where $v \ (> 0)$ is the specific volume, u is the velocity, $\mu \ (> 0)$ is the constant coefficient of viscosity and p is the pressure which satisfies

$$p'(v) < 0, \quad p''(v) > 0 \quad \text{for } v > 0. \tag{2.3}$$

The typical example of p is the isentropic case $p = av^{-\gamma}$ ($a > 0$, $\gamma \geq 1$). We first recall the case $(v_\pm, u_\pm) = (\bar{v}, \bar{u})(\bar{v} > 0, \bar{u} \in R)$. In this case, Kanel [10] showed the global stability of the constant state $(\bar{v}, \bar{u})$, that is, if

$$v_0 - \bar{v}, \quad u_0 - \bar{u} \in H^1, \quad \inf v_0 > 0, \tag{2.4}$$

then the solution of (2.1), (2.2) has a unique global solution (v, u) satisfying

$$
\begin{aligned}
&(v - \bar{v}, u - \bar{u}) \in C^0([0, \infty); H^1), \\
&\sup_{x \in \mathbb{R}} |(v - \bar{v}, u - \bar{u})(t, x)| \to 0 \quad \text{as} \quad t \to \infty.
\end{aligned}
\tag{2.5}
$$

In what follows, let us set $(\bar{v}, \bar{u}) = (1, 0)$ and $p = av^{-\gamma}$ for simplicity. In the proof of this global stability, the crucial *a priori* estimate is the uniform boundedness of v as

$$\exists C > 0 \quad \text{such that} \quad C^{-1} \leq v(t, x) \leq C \quad \text{for } \forall t \geq 0, \forall x \in \mathbb{R}. \tag{2.6}$$

The estimate (2.6) can be obtained by the following basic two energy equalities for (2.1), (2.2):

$$\int \frac{u^2}{2} + a\Psi(v)dx \Big|_0^t + \int_0^t \int \mu \frac{u_x^2}{v} \, dx \, ds = 0, \tag{2.7}$$

where

$$\Psi(v) = \begin{cases} (v - 1) - \log v & \gamma = 1, \\ (v - 1) + (v^{1-\gamma} - 1)/(\gamma - 1), & \gamma > 1 \end{cases}$$

and

$$\int \mu(v_x/v)^2/2 - u(v_x/v)dx \Big|_0^t + \int_0^t \int |p'(v)| v_x^2/v dx ds = \int_0^t \int u_x^2/v \, dx \, ds, \tag{2.8}$$

where (2.7) is derived by multiplying the second equation of (2.1) by u, (2.8) by (v_x/v). In fact, if we define $\Theta(v) = \int_1^v \Psi(s)^{1/2} s^{-1} ds$, it follows from (2.7) and (2.8) that

$$
\begin{aligned}
|\Theta(v(t, x))| &\leq \left| \int_{-\infty}^x \Theta_x dx \right| \leq \int \Psi(v)^{1/2} |v_x|/v dx \\
&\leq \left(\int \Psi(v)dx \right)^{1/2} \left(\int (v_x/v)^2 dx \right)^{1/2} \leq C(v_0, u_0) < \infty.
\end{aligned}
\tag{2.9}
$$

Since $\Theta(v) \to \infty$ (resp. $-\infty$) as $v \to \infty$ (resp. $+0$), we obtain (2.6). Once we have (2.6), it is not difficult to obtain the following *a priori* estimate :

$$\|(v - 1, u)(t)\|_1^2 + \int_0^t \|v_x(s)\|^2 + \|u_x(s)\|_1^2 ds \leq C(v_0, u_0) < \infty. \tag{2.10}$$

By (2.6) and (2.10), we can have the global existence and the asymptotics in the same way as in Section 1.

Remark 2.1. For the full system including the absolute temperature θ, if the gas is ideal and polytropic, Kazhikhov [16] showed by his skillful arguments the global existence of the solution provided $(v_0 - \bar{v}, u_0, \theta_0 - \bar{\theta}) \in H^1$, $\inf v_0 > 0$ and $\inf \theta_0 > 0$. However, the asymptotics toward the constant state $(v, 0, \bar{\theta})$ are not known in general. Kawashima and Nishida [14] showed the asymptotics provided the ratio of specific heat γ is suitably close to one. Furthermore, for a more general full system, Kawashima and Okada [15] showed both the global existence and the asymptotics for small initial data. Further asymptotic properties, asymptotics toward a self-similar solution made through Burger's equation, are investigated in Kawashima [11] and Nishida [33].

Let us turn to the case $(v_-, u_-) \neq (v_+, u_+)$. The large-time behavior of the solution is closely related to that of the Riemann problem for the corresponding inviscid system :

$$v_t - u_t = 0\,,$$
$$u_t + p(v)_x = 0\,, \tag{2.11}$$

with

$$(v, u)(0, x) = (v_0^R, u_0^R)(x) \equiv \begin{cases} (v_-, u_-) & x < 0\,, \\ (v_+, u_+) & x > 0\,, \end{cases} \tag{2.12}$$

In what follows, we put $w = (v, u)$, $w_\pm = (v_\pm, u_\pm)$ ... and so on. For a given state w_- $(v_- > 0$, $u_- \in \mathbb{R})$, in a suitable neighbourhood $\Omega \subset \mathbb{R}_w^2$ of w_-, we define the shock curves $S_i(w_-)$ $(i = 1, 2)$ by

$$S_i(w_-) = \{w \in \Omega; u = u_- - (v - v_-)s_i(v, v_-), \lambda_i(v) \leq \lambda_i(v_-)\}, \tag{2.13}$$

$$(s_1(v, v_-) = -|(p(v) - p(v_-))/(v - v_-)|^{1/2}\,, s_2 = -s_1)$$

and the rarefaction curves $R_i(w_-)$ $(i = 1, 2)$ by

$$R_i(w_-) = \left\{w \in \Omega; u = u_- - \int_{v_-}^v \lambda_i(s)ds, \lambda_i(v) \geq \lambda_i(v_-)\right\}, \tag{2.14}$$

where $\lambda_1(v) = -|p'(v)|^{1/2}$ and $\lambda_2 = -\lambda_1$ are the distinct eigenvalues of the matrix

$$A(v) = \begin{pmatrix} 0 & -1 \\ p'(v) & 0 \end{pmatrix}.$$

We note that S_i is given by the Rankine-Hugoniot condition and the shock condition, and R_i is given as a part of the integral curve of the right eigenvector $r_i(v)$ of $A(v)$. We also define the regions $SS(w_-)$, $RR(w_-)$ and $SR(w_-)$ by

$$SS(w_-) = \{w \in \Omega; u < u_- - (v - v_-)s_i(v, v_-) \text{ for } i = 1, 2\}, \tag{2.15}$$

$$RR(w_-) = \left\{w \in \Omega; u > u_- - \int_{v_-}^v \lambda_i(s)ds \quad \text{for } i = 1, 2\right\}, \tag{2.16}$$

$$SR(w_-) = \Omega \setminus (S_1 \cup S_2 \cup R_1 \cup R_2 \cup SS \cup RR). \tag{2.17}$$

Then the following results are well–known. If $w_+ \in S_1 \cup S_2 \cup SS$, the Riemann problem (2.11), (2.12) has a discontinuous weak solution. More precisely, if $w_+ \in S_i(w_-)$ the solution consists of two constant states $w_\pm$ and a shock discontinuity with the speed $s_i(v_-, v_+)$; if $w_+ \in SS(w_-)$, three constant states $w_\pm$, w^- and two shock discontinuities with the speeds $s_1(v_-, v)^-$ and $s_2(v, v_+^-)$ where w^- is uniquely determined by $w^- \in S_1(w_-)$ and $w_+ \in S_2(\bar{w})$. If $w_+ \in R_1 \cup R_2 \cup RR$, the Riemann problem has a continuous weak solution with the form $w^R(x/t)$. More precisely, if $w_+ \in R_i(w_-)$, the solution $w^R(x/t)$ consists of two constant states $w_\pm$ and a centered rarefaction wave; if $w_+ \in RR$, three constant states $w_\pm$, w^- and two centered rarefaction waves where w^- is determined by $w^- \in R_1(w_-)$ and $w_+ \in R_2(\bar{w})$. We call $w^R(x/t)$ simply "rarefaction wave". Finally, if $w_+ \in SR(w_-)$, then the weak solution consists of both shock wave and rarefaction wave.

2.1 Stability of Viscous Shock Waves

The arguments in this section basically follow from Matsumura-Nishihara [27]. In the case $w_+ \in S_i(w_-)$ ($i = 1$ or 2), there is a corresponding viscous shock wave $W_i(x - st; w_-, w_+)$ which connects w_- and w_+ smoothly where $s = s_i(v_-, v_+)$ in (2.13). In fact, $W_i(\xi) = (V, U)_i(\xi)$ is given as a smooth solution of the problem:

$$
\begin{aligned}
s\mu V_\xi &= V(p(V_\pm) + s^2(v_\pm - V) - p(V)), \quad \xi \in \mathbb{R}, \\
U + sV &= u_\pm + sv_\pm, \quad W(\pm\infty) = w_\pm
\end{aligned}
\tag{2.18}
$$

where $s = s_i(v_-, v_+)$. The convexity of $p(v)$ easily implies that if $w_+ \in S_i(w_-)$, there exists a smooth solution W of (2.18) which is a monotone function and unique up to a shift in ξ. Let W_i be a solution of (2.18) and $w_+ \in (S_1 \cup S_2 \cup SS)(w_-)$. Then we expect the asymptotic state $W^S(t, x)$ of the solution of (2.1), (2.2) is given by

$$
W^S(t, x) = W_i(x - s_i t + \alpha; w_-, w_+), \quad w_+ \in S_i(w_-), \tag{2.19}
$$

$$
\begin{aligned}
W^S(t, x) &= W_1(x - s_1 t + \alpha; w_-, \bar{w}) \\
&\quad + W_2(x - s_2 t + \alpha; \bar{w}, w_+) - \bar{w}, \quad w_+ \in SS(w_-), \tag{2.20}
\end{aligned}
$$

for some shifts α and β. Concerning the shift α in (2.19), we assume

$$
\int w_0(x) - W_i(x)\,dx = \alpha(w_+ - w_-). \tag{2.21}
$$

Since W is a vector, we emphasize that (2.21) does not hold in general, contrary to the scalar case in Section 1.1. Liu [19] proposed that α should be determined by the formula

$$
\int w_0(x) - W_2(x)\,dx = \alpha(w_+ - w_-) + \beta r_1(w_-) \tag{2.22}
$$

for $w_+ \in S_2(w_-)$ (for $w_+ \in S_1$, the corresponding formula holds). It is noted that α and β in (2.22) are always uniquely determined since $(w_+ - w_-)$ and $r_1(w_-)$ are linearly independent. Our assumption (2.21) can be restated as $\beta = 0$ in (2.22). Then we note

$$\int w_0(x) - W_i(x + \alpha)dx = 0, \tag{2.23}$$

and assume

$$\Phi_0(x) = (\varphi_0, \psi_0)(x) \equiv \int_{-\infty}^{x} (w_0(y) - W_i(y + \alpha))\, dy \in L^2. \tag{2.24}$$

For $w_+ \in SS(w_-)$, we can always choose the shifts α and β such that

$$\int w_0(x) - (W_1(x) + W_2(x) - \bar{w})\, dx = \alpha(\bar{w} - w_-) + \beta(w_+ - \bar{w}) \tag{2.25}$$

because $(w^- - w_-)$ and $(w_+ - \bar{w})$ are linearly independent. Then noting that

$$\int w_0(x) - W^S(0, x; \alpha, \beta)\, dx = 0, \tag{2.26}$$

we assume that

$$\Phi_0(x) = (\varphi_0, \psi_0)(x) \equiv \int_{-\infty}^{x} (w_0(y) - W^S(0, y; \alpha, \beta))\, dy \in L^2. \tag{2.27}$$

We further assume for the initial data

$$w_0 - w_0^R \in L^2 \cap L^1, w_{0,x} \in L^2, \tag{2.28}$$

which implies $\Phi_0 \in H^2$. Then, under all assumptions above, we have

Theorem 2.2. *For each $w_+ \in (S_1 \cup S_2 \cup SS)(w_-)$, there exists a positive constant ε_0 such that if $\|\Phi_0\|_2 + |w_+ - w_-| \le \varepsilon_0$, then the problem (2.1), (2.2) has a unique global solution w satisfying*

$$w - W^S \in C^0([0, \infty); H^1) \cap L^2(0, \infty; H^1), u_x \in L^2(0, \infty; H^1), \tag{2.29}$$

$$\sup_{x \in \mathbb{R}} |w(t, x) - W^S(t, x)| \to 0 \quad \text{as } t \to \infty. \tag{2.30}$$

Remark 2.3. (i) In the case $p = av^{-\gamma}$ ($\gamma \geq 1$) and $w_+ \in S_i(w_-)$, Matsumura and Nishihara [27] showed the global existence and asymptotics under the condition (cf.[34])

$$(\gamma - 1)|w_+ - w_-| + \|\Phi_0\|_2 \leq \varepsilon_0 \,. \tag{2.31}$$

Especially, if $\gamma = 1$, (2.31) means any large viscous shock wave is asymptotically stable provided $\int w_0 - W^S \, dx = 0$.

(ii) Kawashima and Matsumura [12] showed similar results for the full system under the condition $\int w_0 - W^S \, dx = 0$.

(iii) Liu [19], [20] reported similar results even for the case $\int w_0 - W^S \, dx \neq 0$ ($\beta \neq 0$ in (2.22)) by his technical characteristic energy method. However, his arguments seem to include a crucial difficulty to be corrected. Just recently, Szepessy and Xin [39] announced to succeed in it.

Here we only show the first step of the *a priori* estimate for Theorem 2.2. Let us assume $w_+ \in S_2(w_-)$ and $|w_+ - w_-| \leq 1$ for simplicity. Then we note that

$$V_\xi(\xi) > 0, \quad v_- < V(\xi) < V_+, \quad |V_\xi(\xi)| \leq C|v_+ - v_-| \,, \tag{2.32}$$
$$\text{for} \quad \xi \in \mathbb{R}, \quad \text{and} \quad s = s_2(v_+, v_-) > 0 \,.$$

As in the previous arguments, we seek the solution w in the form

$$w = W^S + \Phi_x \,, \quad (W^S = (V, U), \Phi = (\varphi, \psi)) \tag{2.33}$$

where Φ satisfies the following "integrated system":

$$\varphi_t - \psi_x = 0 \,,$$
$$\psi_t + (p(V + \varphi_x) - p(V)) = \mu\left(\frac{\psi_{xx}}{V + \varphi_x} + \frac{sV_x\,\varphi_x}{V(V + \varphi_x)}\right) \,, \tag{2.34}$$

$$\Phi(0, x) = \Phi_0(x) \in H^2 \,. \tag{2.35}$$

We rewrite (2.34) in the form

$$\varphi_t - \psi_x = 0 \,,$$
$$\psi_t + p'V\varphi_x - \mu(\frac{\psi_{xx}}{V} + \frac{sV_x\,\varphi_x}{V^2}) = F \,, \tag{2.36}$$

with

$$F = O(|\varphi_x|^2 + |\varphi_x||\psi_{xx}|) \,.$$

Multiplying the second equation of (2.36) by $\frac{\psi}{|p'(V)|}$ and integrating it over $[0, t] \times \mathbb{R}$, we obtain

$$\int \varphi^2 + |p'(V)|\psi^2 \, dx|_0^t + \int_0^t \int sV_x \frac{p''(V)}{|Vp'(V)|^2} \, dx \, ds + 2\int_0^t \int \frac{\mu}{V|p'(V)|}\psi_x^2 \, dx \, ds =$$
$$= 2\int_0^t \int \psi F + \mu\left(\frac{1}{V|p'(V)|}\right)_V V_x\,\psi\psi_x + \mu\frac{sV_x\varphi_x\psi}{V} \, dx \, ds \,. \tag{2.37}$$

115

Now we set $N(t) = \sup_{0 \le s \le t} \|\Phi(s)\|_2$ and assume $N(T) \le 1$. Then making use of (2.32), we can estimate (2.37) as follows

$$\|\Phi(t)\|^2 + \int_0^t (\|V_x^{1/2}\psi(s)\|^2 + \|\psi_x(s)\|^2)\, ds$$
$$\le C \left(\|\Phi_0\|^2 + (N(t) + |w_+ - w_-|) \int_0^t \|\varphi_x\|^2 + |\psi_x\|^2\, ds \right). \tag{2.38}$$

We may multiply the second equation of (2.34) by $-V\varphi_x$ to get the estimate

$$\|\varphi_x(t)\|^2 + \int_0^t \|\varphi_x(s)\|^2\, ds$$
$$\le C \left(\|\Phi_0\|^2 + (N(t) + |w_+ - w_-|) \int_0^t \|\varphi_x\|^2 + |\psi_x\|^2\, ds \right). \tag{2.39}$$

Proceeding these energy estimates, we finally reach

$$\|\Phi(t)\|^2 + \int_0^t \|\varphi_x(s)\|^2\, ds + \|\psi_x(s)\|^2)\, ds$$
$$\le C \left(\|\Phi_0\|^2 + (N(t) + |w_+ - w_-|) \int_0^t \|\varphi_x\|^2 + |\psi_x\|^2\, ds \right). \tag{2.40}$$

Hence, if $N(T)$ and $|w_+ - w_-|$ are suitably small, we can have the desired *a priori* estimate as in Proposition 1.2. The remaining procedure to show the global existence and asymptotics can be done along the same line as in the previous Sections.

2.2 Asymptotics Toward Rarefaction Waves

In this section, we study the case $w_+ \in (R_1 \cup R_2 \cup RR)(w_-)$ according to the arguments in Matsumura-Nishihara [28]. In this case, the solution of (2.1), (2.2) is expected to tend toward the rarefaction wave $w^R(x/t)$. We assume that

$$w_0 - w_0^R \in L^2, \quad w_{0,x} \in L^2. \tag{2.41}$$

Then we have

Theorem 2.4. *For each* $w_- \in (R_1 \cup R_2 \cup RR)(w_-)$, *there exists a positive constant* ε_0 *such that if* $\|w_0 - w_0^R\| + \|w_{0,x}\| + |w_+ - w_-| \le \varepsilon_0$, *then the problem* (2.1), (2.2) *has a unique global solution* w *satisfying*

$$w - w_0^R \in C^0([0,\infty); L^2), \quad w_x \in C^0([0,\infty); L^2) \cap L^2(0,\infty; L^2),$$
$$u_{xx} \in L^2(0,\infty; L^2),$$
$$\sup_{x \in \mathbb{R}} |w(t,x) - w^R(x/t)| \to 0 \quad \text{as} \quad t \to \infty.$$

116

Remark 2.5. Kawashima, Matsumura and Nishihara [13] succeeded in removing the smallness condition for the initial data, and obtained similar results for the full system of ideal and polytropic gas. For a more general system, also refer to Liu and Xin [21] and Xin [43]. Furthermore, Xin [44] treated a problem of zero dissipation limit. Recently, Matsumura and Nishihara [29] showed that if $p(v) = a\,v^{-\gamma}$ $(1 \le \gamma \le 2)$, the solution asymptotically behaves as $w^R(x/t)$ without the smallness conditions of both the initial data and the strength of the rarefaction wave $|w_+ - w_-|$.

Here, according to the arguments in [29], we roughly show how to derive an *a priori* estimate without any smallness conditions in the relatively easier case $p = av^{-1}$. For simplicity, we set $a = 1$ and assume $w_+ \in R_2(w_-)$. Then the rarefaction wave $w^R(x/t)$ is constructed by the formula

$$\lambda_2(v^R(\xi)) = \sigma^R(\xi; \lambda_2(v_-), \lambda_2(v_+)), \qquad u^R(\xi) \in R_2(w_-), \qquad (2.42)$$

where σ^R is given by (1.28). In the same way as in Section 1.2, we construct an approximate function $W = (V, U)$ of w^R by

$$\lambda_2(V(t,x)) = \sigma(t, x; \lambda_2(v_-), \lambda_2(v_+)), \quad U(t,x) \in R_2(w_-), \qquad (2.43)$$

where σ is the solution of (1.31) with the initial data σ_0 of (1.43). It is noted that W is a smooth solution of the inviscid system (2.11) satisfying

$$V_t = U_x > 0, \quad v_+ < V < v_-, \quad |V_x| \le C\,V_t \le C\varepsilon, \qquad (2.44)$$

and similar decay properties as in Lemma 1.6, especially

$$\int_0^\infty \|W_x(t)\|^2 + \|W_{xx}(t)\|\,dt < \infty \qquad (2.45)$$

as in (1.44). If we set $w = W + \Phi$ ($\Phi = (\varphi, \psi)$), the problem (2.1), (2.2) is rewritten in the form

$$\varphi_t - \psi_x = 0,$$

$$\psi_t + (p(v) - p(V))_x - \mu\left(\frac{u_x}{v} - \frac{U_x}{V}\right)_x = \mu\left(\frac{U_x}{V}\right)_x \qquad (2.46)$$

$$\Phi(0, x) = \Phi_0(x) = w_0 - W(0, x) \in H^1. \qquad (2.47)$$

Multiplying the first equation of (2.46) by $(p(V) - p(v))$ and the second equation by ψ, we have

$$\int \frac{1}{2}\psi^2 + (\tilde{v} - 1 - \log \tilde{v})\,dx\Big|_0^t + \int_0^t \int \frac{\mu}{v} Q \quad dx\,ds,$$

$$\le \mu \int_0^t \|\psi\|\|(U_x/V)_x\|ds, \qquad (2.48)$$

where $\tilde{v} = v/V$ and $Q = \psi_x^2 - \psi_x\varphi V_t/V + \varphi^2 V_t/\mu V^2$. By (2.44)–(2.45), we can see $(U_x/V)_x \in L^1(0, \infty; L^2)$ and Q is regarded as a positive definite quadratic form of ψ_x and $\varphi V_t^{1/2}$ for suitably small ε. So it follows from (2.48) that

$$\int \psi^2 + (\tilde{v} - 1 - \log\tilde{v})\,dx + \int_0^t \int \psi_x^2/v + \varphi^2 V_t/v\,dx\,ds \leq C(\Phi_0) < \infty. \quad (2.49)$$

Next we rewrite the second equation of (2.46) in terms of $\tilde{v} = v/V$ as

$$(\mu\frac{\tilde{v}_x}{\tilde{v}} - \psi)_t + \frac{\tilde{v}_x}{\tilde{v}^2 V} + \frac{(1 - \tilde{v})V_x}{\tilde{v}V^2} = -\mu(\frac{U_x}{V})_x. \quad (2.50)$$

Multiplying (2.50) by $\tilde{v}_x/\tilde{v}$, it holds

$$\int \frac{\mu}{2}(\tilde{v}_x/\tilde{v})^2 - \psi(\tilde{v}_x/\tilde{v})\,dx\Big|_0^t + \int_0^t \int \tilde{v}_x^2/\tilde{v}^3 V\,dx\,ds$$

$$= \int_0^t \int \psi_x^2/v - \psi_x\varphi V_t/vV + \varphi V_x\tilde{v}_x/\tilde{v}^2 V^3 - \mu(U_x/V)_x(\tilde{v}_x/\tilde{v})\,dx\,ds. \quad (2.51)$$

If we note (2.44), (2.45) again and the estimate

$$\left|\int \varphi V_x\tilde{v}_x/\tilde{v}^3 V^3\,dx\right| \leq \frac{1}{2}\int \tilde{v}_x^2/\tilde{v}^3 V\,dx + C\int \varphi^2 V_t/v\,dx, \quad (2.52)$$

it is easy to see from (2.49) and (2.51) that

$$\int (\tilde{v}_x/\tilde{v})^2\,dx + \int_0^t \int \tilde{v}_x^2/\tilde{v}^3\,dx\,ds \quad \leq C(\Phi_0) < \infty. \quad (2.53)$$

Once we have (2.49) and (2.53), the same technique as in (2.9) gives the uniform boundedness of $\tilde{v}$, eventually v, from below and above. The remaining process is the same as in the previous arguments.

Some Open Problems. (i) The decay rate of the solution in both Sections 2.1 and 2.2 is not known. We expect similar results as for the scalar equations (cf. Remarks 1.4 and 1.8).

(ii) In the case $w_+ \in SR(w_-)$, both the global existence and the asymptotic behavior are not known entirely. We expect the asymptotic state of the solution is a superposition of both viscous shock wave and rarefaction wave. It seems interesting to consider how the shift of viscous shock wave is determined.

(iii) In the Riemann problem to the full system, it is well-known that there is a case where the weak solution consists of the constant states and a contact discontinuity.

The global existence and the asymptotics of the solution of the corresponding viscous system are not known.

3. Stability of Viscous Shock Waves Without Genuine Nonlinearity

Recently, Matsumura and Kawashima [25] succeeded in obtaining some stability results of viscous shock waves without genuine nonlinearity. To clearly expose their crucial ideas, we consider the case of the scalar problem (1.1), (1.4) again under the conditions

$$f''(u) \gtrless 0 \quad for \quad u \gtrless 0 \text{and} \quad f'''(u) > 0 \quad for \quad u \neq 0, \tag{3.1}$$

instead of the convex condition (1.2). The typical example of (3.1) is $f(u) = u^3$. We note (3.1) implies that f has an inflection point at $u = 0$, therefore equation (1.1) is neither genuinely nonlinear nor linearly degenerate. Even for this case, under the Rankine-Hugoniot condition (1.5) and the shock condition (1.6), there is a viscous shock wave $U(\xi)$ ($\xi = x - st$) connecting u_- and u_+ which is determined by

$$\mu U_\xi = f(U) - sU - (f(u_\pm) - su_\pm) \equiv h(U),$$
$$U(\pm\infty) = u_\pm . \tag{3.2}$$

Since we are interested in the viscous shock wave around the inflection point of f, we here assume that

$$u_- < 0 < u_+. \tag{3.3}$$

Then it follows from (3.2) that

$$U_\xi(\xi) < 0 \quad and \quad u_+ < U(\xi) < u_- \quad for \quad \xi \in \mathbb{R}. \tag{3.4}$$

Under the above assumptions, we can see that entirely the same statements as in Theorem 1.1 hold. In this case, as easily seen in the energy inequality (1.22), we need more efforts to get the *a priori* estimate as (1.23). In fact, since $f''(U)$ changes its sign at $U = 0$, the standard energy estimate (1.22) does not work. Here we try to manipulate a positive weight function $w(U)$ which changes its form depending on the sign of $f''(U)$. Multiplying (1.21) by $w(U)\varphi$, we have

$$\int \frac{1}{2} w(U)\varphi^2 \, dx|_0^t + \int_0^t \int \frac{-1}{2} w(U)z_\varepsilon(U)U_\xi\varphi^2 + (1-\varepsilon)\mu w(U)\varphi_x^2 \, dx \, ds$$
$$\leq \int_0^t \int w(U)\varphi F \, dx \, ds , \tag{3.5}$$

for any $\varepsilon \in (0,1)$, where $z_\varepsilon(U)$ is given by

$$z_\varepsilon(U) = f''(U) - (s - f'(U))\frac{w'(U)}{w(U)} + \frac{h(U)}{2\varepsilon}\left(\frac{w'(U)}{w(U)}\right)^2 . \tag{3.6}$$

Here we have used $\mu U_\xi = h(U)$ and the Schwarz inequality. Inspired by the form of (3.6), we choose $w(U)$ as

$$w(u) = \begin{cases} 1, & u \in [0, u_-], \\ (s - f'(0))^2 (s - f'(u))^{-2}, & u \in [u_+, 0]. \end{cases} \quad (3.7)$$

This $w(U)$ verifies $w(U) \geq 1$ and is a $C^1 - function$ on $[u_+, u_-]$. Then the corresponding z_ε in (3.6) has the form

$$z_\varepsilon(U) = \begin{cases} f''(U), & u \in [0, u_-], \\ -(1 - \gamma(U)/\varepsilon) f''(U), & u \in [u_+, 0], \end{cases} \quad (3.8)$$

where $\gamma(U) = 2h(U)(s - f'(U))^{-2} f''(U)$. By (3.1), it is not hard to see that

$$0 < \bar{\gamma} < 1, \quad \text{where} \quad \bar{\gamma} = \max_{u \in [u_+, 0]} \gamma(u). \quad (3.9)$$

Therefore, choosing $\varepsilon \in (\bar{\gamma}, 1)$ in (3.8), we have $z_\varepsilon(U(\xi)) \geq 0$ for any $\xi \in \mathbb{R}$, and eventually can have the desired estimate as (1.23) from (3.5). Thus, we can prove the statements in Theorem 1.1.

Remark 3.1. We can replace condition (3.1) by

$$f''(u) \lesseqgtr 0 \quad \text{for} \quad u \gtreqless 0 \quad \text{and} \quad f'''(u) < 0 \quad \text{for} \quad u \neq 0, \quad (3.10)$$

under consideration. In fact, if u solves (1.1) with $f(u)$ satisfying (3.10), then the change of independent variables, $y = -x$, transforms (1.1) into

$$u_t + g(u)_y = \mu u_{yy}, \quad (3.11)$$

where $g(u) = -f(u)$, and this $g(u)$ satisfies (3.1). Simple examples of $f(u)$ satisfying either (3.1) or (3.10) are

$$f(u) = \alpha u + \beta u^{2m+1}, \quad (3.12)$$

$$f(u) = \beta u(1 + u^2)^{-1/2}, \quad (3.13)$$

where α, β are constants with $\beta \neq 0$, and $m \geq 1$ is an integer.

Our above arguments are also applicable to the system case. The first application is a model system of viscoelasticity with a non-convex constitutive relation:

$$\begin{aligned} v_t - u_x &= 0, \\ u_t - \sigma(v)_x &= \mu u_{xx}, \end{aligned} \quad (3.14)$$

$$(v, u)(0, x) = (v_0, u_0)(x), \lim_{x \to \pm\infty} (v_0, u_0) = (v_\pm, u_\pm), \tag{3.15}$$

where v denotes the deformation gradient, u the velocity, $\sigma(v)$ the stress, and $\mu > 0$ is the constant coefficient of viscosity, and $\sigma(v)$ is assumed to be a smooth function satisfying the following sign conditions in a suitable neighbourhood of $v = 0$

$$\sigma'(v) > 0, \quad \sigma''(v) \gtrless 0 \quad (v \gtrless 0), \quad \sigma'''(v) > 0 \quad (v \neq 0), \tag{3.16}$$

so that $\sigma(v)$ has an inflection point at $v = 0$. Under assumption (3.16), we can see that the inviscid system (3.14) with $\mu = 0$ is strictly hyperbolic, and all characteristic fields are neither genuinely nonlinear nor linearly degenerate around $v = 0$. Although we omit the details, we can prove the stability of the viscous shock waves, as in the scalar equation, in a neighbourhood of $v = 0$ provided $\int w_0 - W \, dx = 0$ ($w_0 = (v, u)_0$, $W = (U, V)(x + \alpha)$). It seems interesting to discuss the case where $\sigma(v)$ satisfies the sign conditions in a suitable neighbourhood $v = 0$

$$\sigma'(v) > 0, \quad \sigma''(v) \gtrless 0 \quad (v \gtrless 0), \quad \sigma'''(v) < 0 \quad (v \neq 0). \tag{3.17}$$

However, this remains an open question, since we so far could not find a recipe for reducing the case (3.17) to (3.16) as in the scalar equations (cf. Remark 3.1). Next application to the system is a barotropic model system of van der Waals viscous fluid, that is, the problem (2.1), (2.2) with the pressure $p(v)$ given explicitly by

$$p(v) = R\theta/(v - b) - a/v^2 \quad \text{for } v > b, \tag{3.18}$$

where $R > 0$ is the gas constant, $\theta > 0$ the absolute temperature (assumed to be constant), and a and b are positive constants. We are interested in the case where $p(v)$ is strictly decreasing and has just two inflection points in $v > b$. This is the case when $(2/3)^3 b < aR\theta < (3/4)^4 b$. Then it holds $p'(v) < 0$ for $v > b$, and there are two inflection points $\underline{v}$ and $\bar{v}$ of $p(v)$ such that $3b < \bar{v} < 4b < \underline{v}$. Our stability theory is applicable to a viscous shock wave around the first inflection point $v = \underline{v}$, however so far is it not applicable around the second inflection point $v = \bar{v}$.

4. Some Topics from the Initial Boundary Value Problems

In this section, we present some topics related with the initial boundary value problem for a barotropic model of compressible viscous gas with external forces :

$$v_t - u_x = 0,$$
$$u_t + p(v)_x - \mu(\frac{u_x}{v})_x = f(t, \xi), \quad t \geq 0, x \in [0, 1], \tag{4.1}$$
$$(p(v) = av^{-\gamma}, \xi = \int_0^x v(t, y) dy)$$
$$u(t, 0) = u(t, 1) = 0, \quad t \geq 0, \tag{4.2}$$

$$(v, u)(0, x) = (v_0, u_0)(x) \in H^1 \times H_0^1, \inf_{x \in R} v_0(x) > 0, \int_0^1 v_0(x) \, dx = 1, \tag{4.3}$$

121

where $f(t, \xi)$ is the external force which is given as a smooth bounded function of the Euler coordinate

$$(t, \xi) = (t, \int_0^x v(t, y) \, dy).$$ (4.4)

For the problem (4.1)–(4.3), there have been many works on the various topics (please refer to the nice surveys [33] by Nishida and [41] by Valli, and references therein). Among those topics, we first focus our attention on the uniform estimate for the density $\rho \equiv 1/v$

$$0 < C^{-1} \le \rho(t, x) \le C, \quad \forall (t, x) \in [0, \infty) \times [0, 1].$$ (4.5)

It is noted that if $f = 0$, then the estimate (4.5) holds by the same arguments of Kanel as in Section 2. Beirão da Veiga showed in [2] that for any initial data (4.3), if f is suitably small, then the estimate (4.5) holds. Matsumura-Nishida [26] showed that if $\gamma = 1$ then (4.5) holds for any initial data (4.3) and any bounded f. They also showed that if f is T–periodic, then there exists a T–periodic solution for (4.1), (4.2) (the uniqueness of this periodic solution is not known, and several computer simulations rather suggest us a bifurcation phenomenon does occur). Here, according to the arguments in [26], we roughly present a way to have the uniform estimate (4.5) for $\gamma = 1$.

PROOF of the uniform estimate ($\gamma = 1$).: We first note that

$$\int v(t, x) \, dx = \int v_0(x) \, dx = 1.$$ (4.6)

Multiplying the second equation of (4.1) by u and (v/v_x) and combining them (cf.(2.7), (2.8)), we have

$$\frac{d}{dt} E(w(t))^2 + \int \frac{\mu}{2} \left(\frac{u_x^2}{v} + \frac{a \, v_x^2}{v^3} \right) dx = \int uf + \mu \frac{v_x}{2v} f dx,$$ (4.7)

with

$$E(w) = \int \frac{1}{2} u^2 - \frac{\mu}{2} u(v_x/v) + \frac{\mu^2}{4} (v_x/v)^2 + a(v - 1 - \log v) \, dx.$$

Using (4.6), it holds

$$\left| \int uf dx \right| \le \sup |f| \int |u_x| dx$$

$$\le \sup |f| \left(\int u_x^2 / v dx \right)^{1/2},$$

$$\left| \int (v_x/v) f dx \right| \le \sup |f| \left(\int v_x^2 / v^3 dx \right)^{1/2},$$

by the Schwarz inequality. Then equality (4.7) gives

$$\frac{d}{dt}E(w)^2 + \frac{\mu}{4}\int u_x^2/v + av_x^2/v^3\,dx \le C\sup_{t,x}|f|^2. \qquad (4.8)$$

Due to equality (4.6), there exists a point $x_0 = x_0(t) \in [0,1]$ such that $v(t, x_0(t)) = 1$. Therefore, it holds

$$|\log v(x)| \le |\int_{x_0}^x v_x/v\,dx|$$
$$\le \left(\int v_x^2/v^2 dx\right)^{1/2} \le CE(w), \qquad (4.9)$$

or

$$|\log v(x)| \le \left(\int v_x^2/v^3 dx\right)^{1/2}, \qquad (4.10)$$

which implies

$$\exp(-CE(w)) \le \inf_x v \le \sup_x v \le \exp(CE(w)), \qquad (4.11)$$

$$\sup_x |v| \le \exp\left\{\left(\int v_x^2/v^3\,dx\right)^{1/2}\right\}. \qquad (4.12)$$

By (4.11), it is easy to see that $E(w) < \infty$ is equivalent to

$$\|v-1\|^2 + \|v_x\|^2 + \|u\|^2 < \infty \quad \text{and} \quad \inf_x v > 0.$$

Using (4.6) and (4.9), we get

$$\left|\int v - \log v - 1\,dx\right| \le 1 + \int v_x^2/v^3 dx, \qquad (4.13)$$

$$\int u^2 dx \le \int u_x^2/v dx. \qquad (4.14)$$

Especially, by (4.12), $(Y \equiv \int v_x^2/v^3\,dx)$

$$X \equiv \int v_x^2/v^2\,dx \le \sup_x |v| \int v_x^2/v^3\,dx \le Y\exp(Y^{1/2}),$$

which implies

$$G^{-1}(X) \le Y \qquad (4.15)$$

where $G(y)$ is a monotone increasing function defined by

$$G(y) = y \exp(y^{1/2}) \quad (y \geq 0). \tag{4.16}$$

Then estimates (4.13)–(4.15) lead inequality (4.8) to

$$\frac{d}{dt} E(w)^2 + \alpha G^{-1}(C_1 E(w)^2) \leq C_2(1 + \sup_{t,x} |f|^2), \tag{4.17}$$

for some positive constants α, C_1 and C_2. Therefore, noting that G^{-1} is also a monotone increasing function satisfying $G^{-1}(\infty) = \infty$, we can see $E(w(t))$ is uniformly bounded. Thus, (4.11) easily implies the uniform estimate of v, eventually the density ρ (4.5).

Remark 4.1. If the external force is stationary, that is, $f = f(\xi)$, there are more results in connection with the existence and stability of the stationary solutions. In this case, Beirão da Veiga [1] obtained the necessary and sufficient condition, say "Condition A", which assures the existence of the stationary state with positive density satisfying $\int_0^1 v \, dx = 1$. For example, if $f \equiv \lambda$ (constant), the condition

$$|\lambda| < a \left(\frac{\gamma}{\gamma - 1} \right)^\gamma \tag{4.18}$$

gives "Condition A". Under "Condition A", Beiraõ da Veiga [3] showed that the corresponding stationary state is asymptotically stable for suitably small initial perturbations (Matsumura-Padula [31] extended this stability result to multi-dimensional cases with large external potential forces). In particular if $\gamma = 1$, combining the uniform estimate (4.5) in [26], the stationary state turns out to be globally stable for any initial data. Furthermore, Higuchi and Matsumura [7] noted that if $\gamma > 1$ and $\sup_\xi |f(\xi)| \leq a$, then the solution tends toward the stationary state for any initial data (4.3). Comparing with condition (4.18), we expect this stability result to hold under the condition $\sup_\xi |f(\xi)| < a(\gamma/(\gamma - 1))^\gamma$.

Remark 4.2. For the spherically symmetric case, we can also see the corresponding results to Remark 4.1. More precisely, we consider the spherically symmetric motion in a bounded annulus domain $\{x \in \mathbb{R}^3; 0 < r_1 \leq |x| \leq r_2 < \infty\}$ with the Dirichlet zero boundary condition. First, the global existence and uniqueness are proved by Itaya [9]. If the external force $f = 0$, Nagasawa [32] showed the global asymptotic stabiliy of the constant state even for the full system. If the external force is stationary and $\gamma = 1$, by using a skillfull representation of density due to Kazhikov [16] and Nagasawa [32], Matsumura [24] showed the global asymptotic stability of the corresponding stationary state. Since this representation does not work for the time dependent f, the uniform boundedness for the density (eventually, the existence of T–periodic solution) is not known in general, contrary to the usual one-dimensional case. In what follows we assume that the external force is stationary. For $\gamma > 1$, by the arguments in Matsumura-Padula [31], it is easy to see the stationay state to be asymptotically stable under small initial perturbations, as far as f satisfies "Condition A". Higuchi and Matsumura [7] also showed that if $\gamma > 1$ and f is suitably small (depending only on $\int v_0 \, dx$), the stationary state is asymptotically stable for any initial data as long as $\int v_0 \, dx$ is fixed. They further showed for the full system that for any initial data, if f is suitably small the stationary state is asymptotically stable.

Remark 4.3. If "Condition A" fails, the stationary solution includes a vacuum state and looses its regularity at the interface in general. In this case, we rather have to reformulate the Dirichlet problem (4.1)–(4.3) to a free boundary value problem with the density $\rho = 0$ on the free surface. About this problem, there have been several interesting global results by Okada [36] for a barotropic case, Yashima-Benabidallah for a full system and Okada-Makino [37] for a spherically symmetric case.

References

[1] H. Beirão da Veiga, *An L^p − theory for the n − dimensional, stationary, compressible Navier-Stokes equations, and the incompressible limit for compressible fluids. The equilibrium solutions*, Commun. Math. Phys. **109** (1987), 229-248.

[2] H. Beirão da Veiga, *Long time behavior for one-dimensional motion of a general barotropic viscous fluid*, Arch. Rational Mech. Anal. **108** (1989), 141-160.

[3] H. Beirão da Veiga, *The stability of one-dimensional stationary flows of compressible viscous fluids*, Ann. Inst. Henri Poincaré Anal. Non Linéaire **7** (1990), 259-268.

[4] H. F. Yashima, R. Benabidallah, *Equation à symétrie sphérique dún gaz visqueux et calorifère avec la suface libre*, Preprint 2.88(616), Dip. Mat. Pisa, Gennaio.

[5] J. Goodman, *Nonlinear asymptotic stability of viscous shock profiles for conservation laws*, Arch. Rat. Mech. Anal. **95** (1986), 325-344.

[6] Y. Hattori and K. Nishihara, *A note on the stability of the rarefaction wave of the Burgers equation*, Japan J. Appl. Math. **88** (1991), 85-96..

[7] K. Higuchi, A. Matsumura, *Stability of the spherically symmetric stationary flows of compressible viscous gas*, in preparation.

[8] A. M. Il'in, O. A. Oleinik, *Asymptotic behavior of the solutions of the Cauchy problem for certain quasilinear equations for large time*, (Russian), Mat. Sb. **51** (1960), 191-216.

[9] N. Itaya, *On the temporally global solution with radial symmetry of compressible Navier-Stokes equation*, (Japanese), Jimmonronshu of Kobe Univ. Comm. **21**

(1985), 1-10.

[10] Ya. I. Kanel, *On a model system of equations of one-dimensional gas motion*, (Russian), Differencial'nya Uravnenija **4** (1968), 374-380.

[11] S. Kawashima, *Large-time behavior of solutions to hyperbolic parabolic systems of conservation laws and applications*, Proc. Roy. Soc. Edinburgh **106A** (1987), 169-194.

[12] S. Kawashima, A. Matsumura, *Asymptotic stbility of travelling wave solutions of systems for one-dimensional gas motion Commun. Math. Phys.* **101** (1985), 97-127.

[13] S. Kawashima, A. Matsumura and K. Nishihara, *Asymptotic behavior of solutions of a viscous heat-conductive gas*, Proc. Japan Acad. Ser.A **62** (1986), 249-252.

[14] S. Kawashima, T. Nishida, *Global solutions to the initial value problem for the equations of one-dimensional motion of viscous polytropic gas*, J. Math. Kyoto Univ. **21** (1981), 825-837.

[15] S. Kawashima, M. Okada, *Smooth global solutions for the one-dimensional equations in magnetohydrodynamics*, Proc. Japan Acad. Ser. A. **58** (1982), 384-387.

[16] A. V. Kazhikhov, *On the theory of boundary value problems for equations of a one-dimensional non-stationary motion of a viscous heat-conductive gas*, (Russian), Din. Sploshnoi Sredy **50** (1981), 37-62.

[17] A. V. Kazhikhov, V. V. Shelukhin, *The unique solvability "in the large" with respect to time of initial-boundary value problems for one-dimensional equations of a viscous gas*, (Russian), Prikl. Mat. Mech. **41** (1977), 282-291.

[18] P. D. Lax, *Hyperbolic systems of conservation laws, II.*, Commun. Pure Appl. Math. **10** (1957), 537-566.

[19] T. P. Liu, *Nonlinear stability of shock waves for viscous conservation laws*, Memoirs AMS **328** (1985), 1-108.

[20] T. P. Liu, *Shock waves for compressible Navier-Stokes equations are stable*, Commun. Pure Appl. Math. **39** (1986), 565-594.

[21] T. P. Liu, Z. Xin, *Nonlinear stability of rarefaction waves for compressible Navier-Stokes equations*, Commun. Math. Phys. **118** (1988), 451-465.

[22] T. P. Liu, Z. Xin, *Stability of viscous shock waves associated with a system of nonstrictly hyperbolic conservation laws*, to appear in Commun. Pure Appl. Math..

[23] A. Matsumura, *Asymptotics toward rarefaction wave for solutions of the Broadwell model of a discrete velocity gas*, Japan J. Appl. Math. **4** (1987), 489-502.

126

[24] A. Matsumura, *Large-time behavior of the spherically symmetric solutions of an isothermal model of compressible viscous gas*, to appear in Taransport Theory and Statistical Physics.

[25] A. Matsumura, S. Kawashima, *Stability of shock profiles in viscoelasticity with non-convex constitutive relations*, to appear.

[26] A. Matsumura, T. Nishida, *Periodic solutions of a viscous equation*, Recent topics in nonlinear PDE, IV, Lecture Notes in Num. Appl. Anal., 10 (M.Mimura & T.Nishida eds.), Kinokuniya, Tokyo & North-Holland, Amsterdam, 1989, pp. 49-82.

[27] A. Matsumura, K. Nishihara, *On the stability of traveling wave solutions of a one-dimensional model system for compressible viscous gas*, Japan J. Appl. Math. **2** (1985), 17-25.

[28] A. Matsumura, K. Nishihara, *Asymptotics toward the rarefaction waves of the solutions of a one-dimensional model system for compressible viscous gas*, Japan J. Appl. Math. **3** (1986), 1-13.

[29] A. Matsumura, K. Nishihara, *Global stability of the rarefaction wave of a one-dimensional model system for compressible viscous gas*, Commun. Math. Phys. **144** (1992), 325-335.

[30] A. Matsumura, K. Nishihara, *Asymptotics toward the rarefaction wave of the solutions of the Burgers equation with nonlinear degenerate viscosity*, to appear.

[31] A. Matsumura, M. Padula, *Stability of stationary flows of compressible fluids subject to large external potential forces*, Preprint n.168, Dip. Mat. Univ. Studi Ferrara (1992).

[32] T. Nagasawa, *One-dimensional analysis for the motion of compressible viscous heat-conductive fluid*, Doctral thesis of Keio Univ. (1988).

[33] T. Nishida, *Equations of motion of compressible viscous fluids*, Patterns and Waves –Qualitative Analysis of Nonlinear Differential Equations– Studies in Mathematics and its Applications, vol. 18, 1986, pp. 97-128.

[34] T. Nishida, J. A. Smoller, *Solutions in the large for some nonlinear hyperbolic conservation laws*, Commun. Pure Appl. Math. **26** (1973), 183-200.

[35] K. Nishihara, *A note on the stability of traveling wave solutions of Burgers equation*, Japan J. Appl. Math. **2** (1985), 27-35.

[36] M. Okada, *Free boundary value problem for the equation of one-dimensional motion of viscous gas*, Japan J. Appl. Math. **6** (1989), 161-177.

[37] M. Okada, T. Makino, *Free boundary value problem for the equation of spherically symmetric motion of viscous gas*, to appear.

[38] D. H. Sattinger, *On the stability of waves of nonlinear parabolic systems*, Adv. in Math. **22** (1976), 312-355.

[39] A. Szepessy, Z. Xin, *Nonlinear stability of viscous shock waves*, to appear.

[40] A. Valli, *Periodic and stationary solutions for compressible Navier-Stokes equations via a stability method*, Ann. Scuola Norm. Sup. Pisa **(4)10** (1983), 607-647.

[41] A. Valli, *Mathematical results for compressible flows*, Preprint Dipartimento di Matematica, Università degli Studi di Trento, UTM 365 Gennaio (1992), 1-40.

[42] H. F. Weinberger, *Long-time behavior for a regularized scalar conservation law in the absence of genuine nonlinearity*, to appear.

[43] Z. Xin, *Asymptotic stability of rarefaction waves for* 2 × 2 *viscous hyperbolic conservation laws - the two-modes case*, J. Differ. Eqs. **78** (1989), 191-219.

[44] Z. Xin, *Zero dissipation limit to rarefaction waves for the one-dimensional Navier-Stokes equations of compressible isentropic gas*, to appear.

Akitaka Matsumura
Department of Mathematics
Kanazawa University
Kanazawa 920-11
Japan

K.R. RAJAGOPAL
Mechanics of Non–Newtonian Fluids

1. Introduction

This series of four lectures is intended to be an introduction to the mechanics of non–Newtonian fluids. It is meant to merely provide a nodding acquaintance to the burgeoning area of non–Newtonian fluid mechanics. The terminology non-Newtonian fluids can apply to a wide range of materials with widely disparate material structure, the main shared characteristic being the inability of the classical linearly viscous Newtonian model to capture the behavior of these fluids. Polymeric liquids, biological fluids, slurries, suspensions, liquid crystals are but some of the materials which belong to the class of non–Newtonian fluids. .

The departure from Newtonian behavior manifests itself in these materials in many ways. Here we shall discuss the various features exhibited by non–Newtonian fluids. The main points of deviance from Newtonian behavior are:

   (i) *The ability to shear thin or shear thicken,*
  (ii) *The ability to creep,*
 (iii) *The ability to relax stresses,*
 (iv) *The presence of normal stress differences in simple shear flows,*
  (v) *The presence of yield stress.*

A non–Newtonian fluid may possess just one or all of the above characteristics. We shall discuss each of these characteristics briefly.

(i) *Shear thinning or shear thickening*

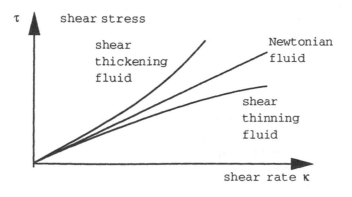

FIGURE 1.

Let us first consider a simple shear flow. In the case of a Newtonian fluid, the relation between the shear stress and shear rate is linear. However, in certain fluids this relationship can be non–linear. We could define a generalized apparent viscosity for these materials, which is the ratio of the shear stress to the shear rate. Materials in which the derivative of the generalized viscosity with respect to the shear rate is negative are called shear thinning fluids. If the derivative of the generalized viscosity to the shear rate is positive, such fluids are called shear thickening fluids. There are examples of both kinds of fluids.

(ii) *Creep*

Consider the material subjected to a step shear stress τ at the time $t = 0$.

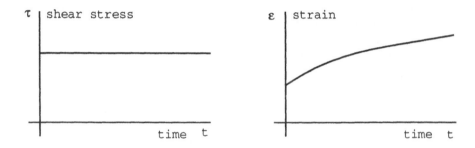

FIGURE 2.

Then, in some non–Newtonian fluids the shear strain would continue to increase as shown. On the removal of the shear stress at time $t = t_0 > 0$, two kinds of behavior are possible as shown in the figures below. If the material is a viscoelastic fluid, then the strain eventually will reach an asymptotic non–zero value. However, if it is a viscoelastic solid the strain will reach an asymptotic value of zero. The continued increase in strain due to a constant applied stress is referred to as creep (see Fig. 3).

(iii) *Stress relaxation*

Next, consider the material subjected to a step strain at time $t = 0$. Then in some materials the stress needed to maintain the strain would decrease with time to zero. Such a material is fluid–like in its behavior. If the stress reduces to a non–zero asymptotic value, the material is solid–like (see Fig.4). The decrease in the value of the stress in order to maintain a constant value of strain is referred to as stress relaxation.

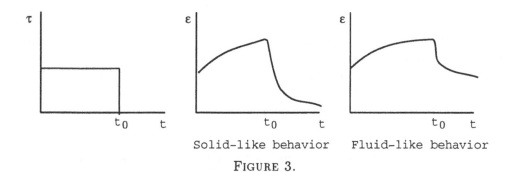

Solid-like behavior   Fluid-like behavior

FIGURE 3.

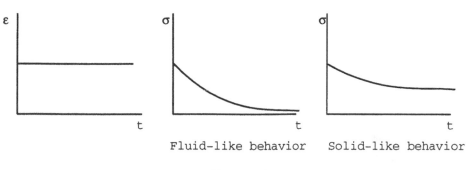

Fluid-like behavior   Solid-like behavior

FIGURE 4.

(iv) *Normal Stress Differences*

Consider simple shear flow of a fluid given by

$$\mathbf{v} = \kappa y \, \mathbf{i},$$

where κ is a constant. In a classical linearly viscous fluid the Cauchy stress $\mathbf{T}$ is given by

$$\mathbf{T} = -p(\rho)\mathbf{I} + \lambda(\operatorname{tr}\mathbf{D})\mathbf{I} + 2\mu\mathbf{D}.$$

It immediately follows that

$$T_{xy} = \mu\kappa,$$
$$T_{xx} - T_{yy} = 0,$$
$$T_{xx} - T_{zz} = 0.$$

However, as we shall see later, there are many materials in which

$$T_{xx} - T_{yy} \neq 0,$$
$$T_{xx} - T_{zz} \neq 0.$$

The fact that the normal stresses T_{xx}, T_{yy} and T_{zz} are different for simple shear flow is referred to as normal stress differences. These normal stress differences are the cause of very many interesting phenomena associated with non–Newtonian fluids like "die swell", "rod–climbing" and secondary flow in pipes of non–circular cross–sections.

(v) *Yield stress*

There are some materials which when subjected to shear stress flow only after a critical value is reached. Such materials are usually referred to as Bingham plastic fluids. The relationship between the shear stress and shear rate is as shown in the figure.

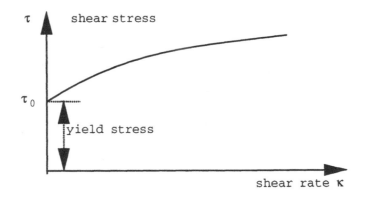

FIGURE 5.

We have discussed very briefly some features associated with non–Newtonian fluids. A reader interested in a more detailed discussion of non–Newtonian fluids is referred to Truesdell and Noll [1], Schowalter [2] and Huilgol [3].

Preliminaries

By a body $\mathcal{B}$ we mean a smooth manifold of dimension three[1]. Its elements $X, Y, \ldots$ are called particles. The configurations $\varphi, \psi, \ldots$ of $\mathcal{B}$ are elements of one to one mappings of $\mathcal{B}$ onto a compact subset of a Euclidean space $\mathcal{E}$ of dimension

---

[1] A more mathematical definition of a body can be found in [1] and [4].

132

three. A motion of the body is a one–parameter family $\{\varphi_t\}$, $-\infty < t < \infty$ of configurations. If φ is a configuration of $\mathcal{B}$, then there is a corresponding non-negative mass density ρ_φ such that

$$m(\mathcal{C}) = \int_{\varphi(\mathcal{C})} \rho_\varphi \, dv \qquad (1.1)$$

for all Borel subsets $\mathcal{C}$ of $\mathcal{B}$, where dv is the Lebesgue volume measure. We shall assume ρ_φ is smooth.

Let $\Omega = \varphi_0(\mathcal{B})$ be a fixed reference placement of $\mathcal{B}$. Since φ_0 is one to one we can identify the body $\mathcal{B}$ with Ω.

Let $\mathbf{X} = \varphi_0(X)$ where $X \in \mathcal{B}$. If $\mathbf{x} = \varphi_t(X)$ denotes the position of particle X at time t, then since φ_0 and φ_t are one to one mappings, the motion of the particle X can be denoted by

$$\mathbf{x} = \chi(\mathbf{X}, t). \qquad (1.2)$$

For any function f defined on $\mathcal{B} \times \mathbb{R}$, there exists a function $\tilde{f}$ defined on $\varphi_0(\mathcal{B}) \times \mathbb{R}$ and $\tilde{\tilde{f}}$ defined on $\varphi_t(\mathcal{B}) \times \mathbb{R}$ such that

$$f(X, t) = \tilde{f}(\mathbf{X}, t) = \tilde{\tilde{f}}(\mathbf{x}, t). \qquad (1.3)$$

If $\Phi : \varphi_0(\mathcal{B}) \times \mathbb{R} \mapsto F_1$ and $\Psi : \varphi_t(\mathcal{B}) \times \mathbb{R} \mapsto F_2$ where F_1 and F_2 are inner product spaces, we use the following notations

$$D_1 \Phi = \nabla \Phi \,, D_2 \Phi = \frac{d\Phi}{dt} = \dot{\Phi} \,,$$
$$D_1 \Psi = \operatorname{grad} \Psi \,, D_2 \Psi = \frac{\partial \Phi}{\partial t} \,, \qquad (1.4)$$

where D_1 and D_2 denote partial derivatives with respect to the first and second variable, respectively. We also denote by

$$\nabla \cdot \Phi = \operatorname{tr}(D_1 \Phi) \,,$$
$$\operatorname{div} \Psi = \operatorname{tr}(D_1 \Psi) \,. \qquad (1.5)$$

The deformation gradient $\mathbf{F}$ associated with the motion is given by

$$\mathbf{F} = \nabla \chi \,. \qquad (1.6)$$

The velocity $\mathbf{v}$ and the velocity gradient $\mathbf{L}$ are defined through

$$\mathbf{v} = \frac{d\chi}{dt} \,, \qquad (1.7)$$
$$\mathbf{L} = \operatorname{grad} \mathbf{v} \,. \qquad (1.8)$$

133

$\mathbf{F} \in \mathrm{Lin}(\mathcal{V}, \mathcal{V})$, where $\mathcal{V}$ is the translational space of $\mathcal{E}$. We shall assume that $\mathbf{F}$ is non–singular. It then follows that

$$\dot{\mathbf{F}}\,\mathbf{F}^{-1} = \mathbf{L}\,. \tag{1.9}$$

Let $\boldsymbol{\xi}$ be the position of $\mathbf{X}$ at time τ. Then, since we assume the motion to be invertible

$$\boldsymbol{\xi} = \chi(\mathbf{X}, \tau) = \chi(\chi^{-1}(\mathbf{x}, t), \tau) = \chi_t(\mathbf{x}, \tau)\,. \tag{1.10}$$

χ_t is called the relative motion. The relative deformation gradient $\mathbf{F}_t(\tau)$ is defined through

$$\mathbf{F}_t(\tau) = \frac{\partial \boldsymbol{\xi}}{\partial \mathbf{x}} = \mathrm{grad}\,\boldsymbol{\xi}\,. \tag{1.11}$$

The relative stretch tensor $\mathbf{C}_t(\tau)$ is given by

$$\mathbf{C}_t(\tau) = \mathbf{F}_t^T(\tau)\mathbf{F}_t(\tau)\,. \tag{1.12}$$

We next define the Rivlin–Ericksen tensors $\mathbf{A}_n$ through (cf. Rivlin and Ericksen [5])

$$\mathbf{A}_n :\equiv \left[\frac{\partial^n \mathbf{C}_t(\lambda)}{\partial \lambda^n}\right]\bigg|_{\lambda=t}\,. \tag{1.13}$$

It follows that

$$\mathbf{A}_1 = (\mathrm{grad}\,\mathbf{v}) + (\mathrm{grad}\,\mathbf{v})^T\,, \tag{1.14}_1$$

and

$$\mathbf{A}_n = \dot{\mathbf{A}}_{n-1} + \mathbf{A}_{n-1}\mathbf{L} + \mathbf{L}^T\mathbf{A}_{n-1}\,. \tag{1.14}_2$$

By the Polar decomposition theorem

$$\mathbf{F}_t(\tau) = \mathbf{R}_t(\tau)\,\mathbf{U}_t(\tau) = \mathbf{V}_t(\tau)\mathbf{R}_t(\tau)\,, \tag{1.15}$$

where $\mathbf{R}_t(\tau)$ is an orthogonal transformation and $\mathbf{U}_t(\tau)$ and $\mathbf{V}_t(\tau)$ are positive definite symmetric tensors. By (1.12) and (1.15),

$$\mathbf{C}_t(\tau) = \mathbf{U}_t^2(\tau)\,. \tag{1.16}$$

$\mathbf{U}_t(\tau)$ and $\mathbf{V}_t(\tau)$ are called stretch tensors.

Let $\varphi(t)$ be any scalar, vector or tensor valued function of time. The history of φ denoted by $\varphi^t(s)$ is defined through

$$\varphi^t(s) = \varphi(t-s)\,, \quad s \geq 0\,. \tag{1.17}$$

134

We are now in a position to define a simple fluid (cf. Noll [6]). We say a fluid is a simple fluid if the Cauchy stress $\mathbf{T}$ is determined by

$$\mathbf{T} = \mathop{\mathfrak{F}}_{s=0}^{\infty} \left[\mathbf{F}_t^t(s), \rho^t(s) \right] . \tag{1.18}$$

We shall be interested in incompressible simple fluids. In this case

$$\mathbf{T} = -p\mathbf{I} + \mathop{\mathfrak{G}}_{s=0}^{\infty} \left[\mathbf{F}_t^t(s) \right] . \tag{1.19}$$

with $\mathop{\mathfrak{F}}_{s=0}^{\infty} [\mathbf{1}^t(s)] = \mathbf{0}$, $\mathbf{1}^t$ being the history which has the constant value of $\mathbf{1}$, $\mathop{\mathfrak{G}}_{s=0}^{\infty}$ and $\mathop{\mathfrak{F}}_{s=0}^{\infty}$ being functionals.

Principle of frame indifference places severe restrictions on the forms of $\mathop{\mathfrak{G}}_{s=0}^{\infty}$. It follows (cf. Truesdell [4]) that

$$\mathbf{T} = -p\mathbf{I} + \mathop{\mathbf{h}}_{s=0}^{\infty} \left[\mathbf{C}_t^t(s) \right] , \tag{1.20}$$

with

$$\mathop{\mathbf{h}}_{s=0}^{\infty} \left[\mathbf{Q}\mathbf{C}_t^t(s)\mathbf{Q}^T \right] = \mathbf{Q} \mathop{\mathbf{h}}_{s=0}^{\infty} \left[\mathbf{C}_t^t(s) \right] \mathbf{Q}^T \tag{1.21}$$

for all $\mathbf{Q}$ that are orthogonal.

In their paper on an approximation theorem for functionals, Coleman and Noll [7] show that for a special class of flows, called retarded flows, the representation for the Cauchy stress of a simple fluid takes on a special form. This follows under appropriate smoothness assumptions on $\mathbf{h}$. We shall not get into a detailed discussion of this important work here but refer the reader to [7]. However, we would like to comment that the study of Coleman and Noll [7] does not provide a hierarchy of constitutive equations as it is usually interpreted. Such an interpretation has led to unnecessary confusion and controversy in the field of non–Newtonian fluid mechanics. These and related issues are discussed in great detail in the recent paper by Dunn and Rajagopal [8].

We conclude this section with a discussion of special flows. In these flows, the stress $\mathbf{T}$ takes on a very simple structure.

The class of flows which we shall discuss are those in which the history of the stretch tensor $\mathbf{C}_t(t-s)$ has constant eigenvalues for all s, the eigen directions possibly changing with s. Such flows are called motions with constant principal relative stretch history. The following result due to Noll [9] provides necessary and sufficient conditions for a flow to be a motion with constant principal relative stretch history.

A flow is a motion with constant relative principal stretch history if and only if there exists an orthogonal tensor $\mathbf{Q}(t)$ and a tensor $\mathbf{M}$ such that

$$\mathbf{F}_0(\tau) = \mathbf{Q}(\tau) e^{\tau\mathbf{M}} , \tag{1.22}$$

135

with $\mathbf{Q}(0) = \mathbf{1}$.

A sufficient condition for a motion with constant relative principal stretch history is (cf. Huilgol [10])

$$\frac{d\mathbf{L}}{dt} = \mathbf{0}. \tag{1.23}$$

Most of the simple flows one encounters in a first course in fluid dynamics like plane Poiseille flow, Coutte flow, Helical flow etc. are all motions with constant principal relative stretch history. Recently, Rajagopal [11] showed that the flow that takes place in the orthogonal rheometer also belongs to the above class. We shall study this flow in some detail in Chapter 4.

In a motion with constant principal relative stretch history the stress takes on a simple form. Wang [12] has shown that the functional reduces to a simple structure and

$$\mathbf{T} = -p\,\mathbf{I} + \mathbf{f}(\mathbf{A}_1, \mathbf{A}_2, \mathbf{A}_3), \tag{1.24}$$

where $\mathbf{A}_1$ and $\mathbf{A}_2$ and $\mathbf{A}_3$ are the first three Rivlin–Ericksen tensors at time t, and $\mathbf{f}$ is a tensor–valued function.

The special subclass of flows with constant principal relative stretch history in which the representation (1.24) is such that

$$\mathbf{M}^2 = \mathbf{0} \tag{1.25}$$

is called Viscometric flows. It can be shown that the condition (1.25) is equivalent to the requirement that

$$\mathbf{A}_n = \mathbf{0}, \quad n \geq 3. \tag{1.26}$$

We shall see later that this implies that a Viscometric flow of a Simple fluid cannot be distinguished from a fluid of complexity two. While in a Viscometric flow the stress is determined by A_1 and A_2, the converse is not true, that is, if the stress is completely determined by $\mathbf{A}_1$ and $\mathbf{A}_2$ it does not imply that the flow is Viscometric. A simple example is the flow in a orthogonal rheometer studied by Rajagopal [11].

2. Fluids of the differential type

A detailed and comprehensive critique of the thermodynamics of fluids of the differential type can be found in the recent work of Dunn and Rajagopal [8]. Results regarding boundedness, asymptotic stability, and uniqueness of flows of such fluids can be found in [13] and [15]. Here, we shall be content with presenting some of the salient results pertinent to the thermomechanics of such fluids. Due to paucity of space we shall not provide details of proofs of the theorems. The interested reader can find them in [13], [14] and [15].

An incompressible homogeneous fluid of complexity n is characterized by four constitutive functions ε (specific internal energy), η (specific entropy), $\mathbf{T}$ (Cauchy stress tensor), and $\mathbf{q}$ (heat flux vector) through

$$\varepsilon = \hat{\varepsilon}(\theta, \operatorname{grad}\theta, \mathbf{L}, \dot{\mathbf{L}}, \ldots, \overset{(n-1)}{\mathbf{L}}),$$

$$\eta = \hat{\eta}(\theta, \operatorname{grad}\theta, \mathbf{L}, \dot{\mathbf{L}}, \ldots, \overset{(n-1)}{\mathbf{L}}),$$

$$\mathbf{T} = -p\mathbf{I} + \hat{\mathbf{T}}(\theta, \operatorname{grad}\theta, \mathbf{L}, \dot{\mathbf{L}}, \ldots, \overset{(n-1)}{\mathbf{L}}),$$

$$\mathbf{q} = \hat{\mathbf{q}}(\theta, \operatorname{grad}\theta, \mathbf{L}, \dot{\mathbf{L}}, \ldots, \overset{(n-1)}{\mathbf{L}}),$$

$$(2.1)_{1-4}$$

where the term $-p\mathbf{I}$ is due to the constraint of incompressibility and $\overset{(n-1)}{\mathbf{L}}$ denotes $(n-1)^{\text{th}}$ time derivatives of $\mathbf{L}$.

We now state without proof the following

Theorem 1. *The response functions $\hat{\varepsilon}$, $\hat{\eta}$, $\hat{\mathbf{T}}$ and $\hat{\mathbf{q}}$ in an incompressible Rivlin–Ericksen fluid of complexity n are compatible with thermodynamics if and only if*

$$\psi = \hat{\psi}(\theta, \mathbf{L}, \dot{\mathbf{L}}, \ldots, \overset{(n-2)}{\mathbf{L}}), \tag{2.2}$$

$$\eta = -\hat{\psi}_\theta \tag{2.3}$$

and

$$\rho\left\{\hat{\psi}_\mathbf{L} \cdot \dot{\mathbf{L}} + \hat{\psi}_{\dot{\mathbf{L}}} \cdot \ddot{\mathbf{L}} + \cdots + \hat{\psi}_{\underset{\mathbf{L}}{(n-2)}} \cdot \overset{(n-1)}{\mathbf{L}}\right\} - \hat{\mathbf{T}} \cdot \mathbf{L} + \frac{\hat{\mathbf{q}} \cdot \operatorname{grad}\theta}{\theta} \leq 0 \tag{2.4}$$

in every thermodynamic process.

Thus, thermodynamics places restrictions on the form of constitutive equations that are allowable. We next state a theorem that asserts that the specific Helmholtz free energy ψ attains a "stationary value" when $\mathbf{L} = \dot{\mathbf{L}} = \cdots = \overset{(n-2)}{\mathbf{L}} = 0$.

Theorem 2. *The free energy of an incompressible Rivlin–Ericksen fluid of complexity n attains a stationary value when*

$$\mathbf{L} = \dot{\mathbf{L}} = \cdots = \overset{(n-2)}{\mathbf{L}} = 0.$$

Prompted by physical expectations, we shall assume that the stationary value is a minimum. Later, this assumption will play a crucial role in determining the sign of material coefficients of fluids of grades two and three. We shall see that requiring that the stationary value be a minimum leads to signs for material coefficients that imply desirable stability characteristics for these fluids.

Assumption 1.

$$\hat{\psi}(\theta, \mathbf{0}, \mathbf{0}, \ldots, \mathbf{0}) \leq \hat{\psi}(\theta, \mathbf{L}, \dot{\mathbf{L}}, \ldots, \overset{(n-2)}{\mathbf{L}}), \tag{2.5}$$

for all $\mathbf{L}, \dot{\mathbf{L}}, \ldots, \overset{(n-2)}{\mathbf{L}} \in \mathcal{J}_0$, where $\mathcal{J}_0 \equiv \{\mathbf{T} \in \text{Lin}(\mathcal{V}, \mathcal{V}) \mid \text{tr}\,\mathbf{T} = 0\}$ is the set of traceless tensors.

We shall also assume that the specific heat ε_θ is strictly positive. This is tantamount to

Assumption 2.

$$\hat{\psi}_{\theta\theta}(\theta, \mathbf{L}, \dot{\mathbf{L}}, \ldots, \overset{(n-2)}{\mathbf{L}}) < 0. \tag{2.6}$$

We shall say that a body is mechanically isolated (cf. [16], [17]) at any instant t if

$$\int_{\partial\Omega_t} \mathbf{T}\mathbf{n} \cdot \mathbf{v}\, da + \int_{\Omega_t} \rho \mathbf{b} \cdot \mathbf{v}\, dv = 0. \tag{2.7}$$

We shall say that the body is immersed in a "thermally passive environment" at temperature θ^0 if (cf. [16], [17])[2]

$$\{\theta(\mathbf{x}, t) - \theta^0\}\, \mathbf{q}(\mathbf{x}, t) \cdot \mathbf{n}(\mathbf{x}, t) \geq 0 \qquad \forall \mathbf{x} \in \partial\Omega_t, \tag{2.8$_1$}$$

$$\{\theta(\mathbf{x}, t) - \theta^0\}\, r(\mathbf{x}, t) \leq 0 \qquad \forall \mathbf{x} \in \partial\Omega_t. \tag{2.8$_2$}$$

A simple example of a "mechanically isolated" body is one in which $\mathbf{v} = \mathbf{0}$ on the boundary of the body, and $\mathbf{b}$ is a conservative body force field. A body in which at any time t heat is not being conducted into (out of) the body at points on the surface where the temperature θ is greater than (less than) the environmental temperature θ^0, and in which heat is not being radiated into (out of) the body at points in the body where the temperature θ is greater than the environmental temperature θ^0 is in a thermally passive environment.

The class of models we are working with is so general that we cannot expect to establish sharp results regarding uniqueness of flows, existence or stability. Even so, we shall see that for this large class of materials we can say a surprising lot about issues of uniqueness and stability.

Let us now define a function $\Phi : \mathbb{R} \mapsto \mathbb{R}$ such that

$$\Phi(t) = \int_{\Omega_t} \rho\left(\varepsilon - \theta^0\eta + \tfrac{1}{2}\dot{\mathbf{x}} \cdot \dot{\mathbf{x}}\right) dv. \tag{2.9}$$

$\Phi(t)$ is called the canonical free energy of the system. If the body is mechanically isolated and undergoes a process that is consistent with a thermally passive environment, it can be shown that

$$\frac{d\Phi}{dt} \leq 0. \tag{2.10}$$

---

[2] As usual, r denotes the rate of external heat communication to the body.

However, since we do not know if $\Phi(0)$ is positive the result is even weaker than it seems.

Due to frame indifference, it follows that

$$
\begin{aligned}
\varepsilon &= \hat{\varepsilon}(\theta, \mathbf{L}, \dot{\mathbf{L}}, \ldots, \overset{(n-2)}{\mathbf{L}}) = \tilde{\varepsilon}(\theta, \mathbf{A}_1, \ldots, \mathbf{A}_{n-1}), \\
\eta &= \hat{\eta}(\theta, \mathbf{L}, \dot{\mathbf{L}}, \ldots, \overset{(n-2)}{\mathbf{L}}) = \tilde{\eta}(\theta, \mathbf{A}_1, \ldots, \mathbf{A}_{n-1}), \\
\mathbf{T} &= -p\mathbf{I} + \hat{\mathbf{T}}(\theta, \mathbf{L}, \dot{\mathbf{L}}, \ldots, \overset{(n-1)}{\mathbf{L}}) = -p\mathbf{I} + \tilde{\mathbf{T}}(\theta, \mathbf{A}_1, \ldots, \mathbf{A}_n), \\
\mathbf{q} &= \hat{\mathbf{q}}(\theta, \mathbf{L}, \dot{\mathbf{L}}, \ldots, \overset{(n-1)}{\mathbf{L}}) = \tilde{\mathbf{q}}(\theta, \mathbf{A}_1, \ldots, \mathbf{A}_n).
\end{aligned}
\tag{2.11}_{1-4}
$$

We now state a theorem without proof concerning the uniqueness of the rest state of Rivlin–Ericksen fluids of complexity n. Proofs of this theorem and the following theorems in this chapter can be found in [13] and [15]. For the sake of brevity we shall skip them here.

Theorem 3. *Let an incompressible Rivlin–Ericksen fluid of complexity n undergo a process that is both mechanically isolated and consistent with a thermally passive environment at a constant temperature θ^0. Further, let*

$$
\dot{\mathbf{x}}(X,0) = \ddot{\mathbf{x}}(X,0) = \cdots = \overset{(n-1)}{\mathbf{a}}(X,0) = 0 \quad \forall X \in \Omega,
\tag{2.12}
$$

where $\overset{(n-1)}{\mathbf{a}}$ is the $(n-1)^{\mathrm{th}}$ acceleration, and let

$$
\theta(X,0) = \theta^0 \quad \forall X \in \Omega.
\tag{2.13}
$$

Then for all $X \in \Omega$ and $t \in [0, \infty)$

$$
\theta(X,t) = \theta^0,
$$

$$
\dot{\mathbf{x}}(X,t) = \ddot{\mathbf{x}}(X,t) = \cdots = \overset{(n-1)}{\mathbf{a}}(X,t) = 0,
\tag{2.14}_{1,2}
$$

and

$$
\mathbf{T}(X,t) = -\bar{p}\mathbf{I}, \quad \mathbf{q}(X,t) = 0,
\tag{2.14}_3
$$

where

$$
\bar{p} = p - \frac{1}{3} \operatorname{tr} \hat{\mathbf{T}}(\theta, \mathbf{0}, \ldots, \mathbf{0}).
\tag{2.15}
$$

Next we shall state without proof a boundedness result regarding flows of Rivlin–Ericksen fluids of complexity n.

Theorem 4. *Suppose an incompressible fluid of complexity n is undergoing a thermodynamic process which is both mechanically isolated and consistent with a thermally passive environment at constant temperature θ^0. Let t' be any particular instant during this process and let us define*

$$\omega(t') = \int_{\Omega_{t'}} \rho \left([\tilde{\psi}(\theta^0, \mathbf{A}_1(t'), \ldots, \mathbf{A}_{n-1}(t')) - \tilde{\psi}(\theta^0, \mathbf{0}, \ldots, \mathbf{0})] \right.$$

$$\left. + \tilde{K}(t')(\theta(t') - \theta^0)^2 + \tfrac{1}{2}\dot{\mathbf{x}}(t') \cdot \dot{\mathbf{x}}(t') \right) dv. \qquad (2.16)$$

Then for all $t \geq t'$

$$0 \leq \int_{\Omega_{t'}} \rho[\tilde{\psi}(\theta^0, \mathbf{A}_1(t), \ldots, \mathbf{A}_{n-1}(t)) - \tilde{\psi}(\theta^0, \mathbf{0}, \ldots, \mathbf{0})]\, dv \ \leq \ \omega(t'),$$

$$0 \leq \int_{\Omega_{t'}} \rho \tilde{K}(t)(\theta(t) - \theta^0)^2\, dv \ \leq \ \omega(t'),$$

$$0 \leq \int_{\Omega_{t'}} \rho\left(\tfrac{1}{2}\dot{\mathbf{x}}(t) \cdot \dot{\mathbf{x}}(t)\right) dv \ \leq \ \omega(t'), \qquad (2.17)$$

where $\tilde{K}(\tau) \equiv \tilde{K}(\theta(x,\tau), \theta^0, \mathbf{A}_1(x,\tau), \ldots, \mathbf{A}_{n-1}(x,\tau))$.

In order to obtain better bounds on the stability of these fluids we make a stricter assumption on the nature of the free energy. Thus far we have assumed that the free energy is a minimum in equilibrium (i.e., when locally in rest). Let us henceforth assume that the free energy is also twice differentiable and that $\tilde{\psi}(\theta, \cdot, \ldots, \cdot)$ is everywhere convex in[3] $\mathcal{J}_o^S \times \cdots \times \mathcal{J}_o^S$, that is

$$\tilde{\psi}_{\mathbf{A}_i \mathbf{A}_j}(\theta, \mathbf{A}_1, \ldots, \mathbf{A}_{n-1}) \cdot (\mathbf{B}_i \otimes \mathbf{B}_j) \geq 0 \qquad (2.18)$$

for all $\mathbf{A}_1, \mathbf{A}_2, \ldots \mathbf{A}_{n-1}$ and $\mathbf{B}_1, \mathbf{B}_2, \ldots \mathbf{B}_{n-1} \in \mathcal{J}_o^S$, where the common indices i and j are summed from 1 to $n-1$. We first establish the following result.

Theorem 5. *Let the hypothesis of Theorem 4 hold, and suppose that $\tilde{\psi}(\theta, \mathbf{0}, \ldots, \mathbf{0})$ is a strict minimum. If $\omega(t')$ is as defined in (2.16), then for any $\delta > 0$, there exists a positive number $N(\delta)$, which depends on δ and the structure of $\tilde{\psi}$ such that for all $t > t'$,*

$$0 \leq \int_{\Omega_t} \rho|\mathbf{A}_i(t)|\, dv \ \leq \ M\,\delta + N(\delta)\omega(t') \quad \forall i = 1, 2, \ldots, n-1, \qquad (2.19)_1$$

$$0 \leq \int_{\Omega_t} \rho\tilde{K} \ \leq \ M\,\delta + N(\delta)\omega(t'), \qquad (2.19)_2$$

---

[3] $\mathcal{J}_0^S \equiv \{\mathbf{T} \in \mathcal{J}_0 \,|\, \mathbf{T} = \mathbf{T}^T\}$

and

$$0 \le \int_{\Omega_t} \frac{1}{2} \rho \dot{\mathbf{x}}(t) \cdot \dot{\mathbf{x}}(t) \, dv \le \omega(t'), \qquad (2.19)_3$$

where

$$M \equiv \int_{\Omega_t} \rho \, dv \, .$$

We have thus far assumed that $\tilde{\psi}(\theta, \cdot, \ldots, \cdot)$ is everywhere convex in $\mathcal{J}_o^S \overbrace{\times \cdots \times}^{(n-1)} \mathcal{J}_o^S$ for all $\theta \in \mathbb{R}^+$. Let us now assume that $\tilde{\psi}(\cdot, \mathbf{A}_1, \ldots, \mathbf{A}_{n-1})$ is concave everywhere in $\mathbb{R}^+$. Then, we can prove the following

Theorem 6. *Let us suppose that* $\varepsilon_\theta > 0$. *Then*

$$\varepsilon = \bar{\varepsilon}(\eta, \mathbf{A}_1, \ldots, \mathbf{A}_{n-1}) = \tilde{\varepsilon}(\bar{\theta}(\eta, \mathbf{A}_1, \ldots, \mathbf{A}_{n-1}), \mathbf{A}_1, \ldots, \mathbf{A}_{n-1}) \qquad (2.20)$$

attains a minimum at $(\eta, 0, \ldots, 0)$, *i.e.,*

$$\bar{\varepsilon}(\eta, \mathbf{A}_1, \ldots, \mathbf{A}_{n-1}) \ge \bar{\varepsilon}(\eta, 0, \ldots, 0), \qquad (2.21)$$

and the temperature relation

$$\bar{\varepsilon}_\eta(\eta, \mathbf{A}_1, \ldots, \mathbf{A}_{n-1}) = \theta \qquad (2.22)$$

holds.
If

$$\tilde{\psi}_{\mathbf{A}_i \mathbf{A}_j}(\theta, \mathbf{A}_1, \ldots, \mathbf{A}_{n-1}) \cdot (\mathbf{B}_i \otimes \mathbf{B}_j) \ge 0 \qquad (2.23)$$

for all $\mathbf{A}_1$, $\mathbf{A}_2$, ... $\mathbf{A}_{n-1}$ *and* $\mathbf{B}_1$, $\mathbf{B}_2$, ... $\mathbf{B}_{n-1} \in \mathcal{J}_o^S$, *then* $\bar{\varepsilon}$ *is convex on* $\mathbb{R}^+ \times \mathcal{J}_o^S \overbrace{\times \cdots \times}^{(n-1)} \mathcal{J}_o^S$.

We now state the following boundedness results on $\bar{\varepsilon}$ and η, which are very similar to those obtained in theorem 5.

Theorem 7. *Let an incompressible Rivlin–Ericksen fluid of complexity* n *undergo a thermodynamic process which is both mechanically isolated and consistent with a thermally passive environment at constant temperature* θ^0. *Let us suppose that* $\bar{\varepsilon}$ *is convex and has a strict minimum in equilibrium. Let* t' *be any instant during this process and let* $\omega(t')$ *be the positive number defined through*

$$\omega(t') = \int_{\Omega_{t'}} \rho \left([\bar{\varepsilon}(\eta(t'), \mathbf{A}_1(t'), \ldots, \mathbf{A}_{n-1}(t')) - \varepsilon^0 - \theta^0(\eta(t') - \eta^0) \right.$$

$$\left. + \tfrac{1}{2} \mathbf{x}(t') \cdot \dot{\mathbf{x}}(t') \right) \, dv \, .$$

$$(2.24)$$

where $\eta^0 \equiv -\tilde{\psi}_\theta(\theta^0, \mathbf{0}, \ldots, \mathbf{0})$ and $\varepsilon^0 \equiv \bar{\varepsilon}(\theta^0, \mathbf{0}, \ldots, \mathbf{0})$. Then for any $\delta > 0$ and all $t \geq t'$ one obtains

$$0 \leq \int_{\Omega_t} \rho |\bar{\varepsilon}(x,t) - \varepsilon^0| \, dv \; \leq \; M\,\delta + (N(\delta) + 1)\,\omega(t'),$$

$$0 \leq \theta^0 \int_{\Omega_t} \rho |\bar{\eta}(x,t) - \eta^0| \, dv \leq M\,\delta + N(\delta)\,\omega(t'),$$

$$\text{(2.25)}_{1-4}$$

$$0 \leq \int_{\Omega_t} \rho |\mathbf{A}_i|(x,t) \, dv \; \leq \; M\,\delta + N(\delta)\,\omega(t'), \quad (i = 1, 2, \ldots, n-1),$$

$$0 \leq \int_{\Omega_t} \tfrac{1}{2}\rho\, \dot{\mathbf{x}}(t) \cdot \dot{\mathbf{x}}(t) \, dv \; \leq \; \omega(t'),$$

where M is the mass of the fluid and $N(\delta)$ is a positive number depending only on δ and the structure of $\bar{\varepsilon}$.

Results regarding boundedness of the kinematic and thermal fields can be established under other assumptions regarding the fields. We refer the reader to [13] and [15] for the same.

3. Fluids of grade three

In this chapter we shall discuss the thermomechanics of a special subclass of fluids of complexity n, namely fluids of grade n (cf. Truesdell and Noll [1]). As the constitutive representation is quite specific we can obtain more precise results regarding the stability of flows of such fluids. Here, we shall confine our discussion to fluids of grade three, which contains as a special subclass fluids of grade two and fluids of grade one (the classical linearly viscous fluids). A more detailed study of the thermomechanics of fluids of grade three can be found in [13] and [15].

We shall confine our attention to isothermal problems and hence shall concern ourselves with purely the mechanical behavior of fluids of grade three. However, before doing this we shall just document a result regarding the thermodynamics of fluids of grade three that has far reaching significance with regard to issues of stability.

The Cauchy stress for a fluid of grade three has the representation

$$\mathbf{T} = -p\,\mathbf{I} + \mu\mathbf{A}_1 + \alpha_1\mathbf{A}_2 + \alpha_2\mathbf{A}_1^2 + \beta_1\mathbf{A}_3 + \beta_2[\mathbf{A}_1\mathbf{A}_2 + \mathbf{A}_2\mathbf{A}_1] + \beta_3(\operatorname{tr}\mathbf{A}_1^2)\mathbf{A}_1 \,. \quad (3.1)$$

We shall now state the following

Theorem 8. *The response functions $\hat{\varepsilon}$, $\hat{\eta}$, $\hat{\mathbf{T}}$ and $\hat{\mathbf{q}}$ of a fluid of grade three are compatible with thermodynamics in the sense that all processes meet the Clausius–Duhem inequality and the assumption that the specific Helmholtz free energy $\hat{\psi}$ meets*

$$\hat{\psi}(\theta, \mathbf{L}, \dot{\mathbf{L}}, \ddot{\mathbf{L}}) \leq \hat{\psi}(\theta, \mathbf{0}, \mathbf{0}, \mathbf{0}) \qquad (3.2)$$

if and only if
(i) the viscosity $\mu(\theta)$ is non–negative, i.e.

$$\mu(\theta) \geq 0, \tag{3.3}$$

(ii) the normal stress coefficients $\alpha_1(\theta)$ and $\alpha_2(\theta)$ meet

$$\alpha_1(\theta) \geq 0, \tag{3.4}$$

$$-\sqrt{24\,\mu(\theta)\,\beta_3(\theta)} \leq \alpha_1(\theta) + \alpha_2(\theta) \leq \sqrt{24\,\mu(\theta)\,\beta_3(\theta)}, \tag{3.5}$$

(iii)

$$\beta_1(\theta) = 0, \quad \beta_2(\theta) = 0, \quad \beta_3(\theta) \geq 0, \tag{3.6}$$

and
(iv) the free energy $\hat{\psi}$ has the following explicit form:

$$\psi = \hat{\psi}(\theta, \mathbf{L}) = \hat{\psi}(\theta, \mathbf{0}) + \frac{\alpha_1(\theta)}{4\rho}\,|\mathbf{A}_1|^2\,. \tag{3.7}$$

Henceforth, we shall dispense with thermal effects and assume all the material moduli to be constant. Thus, in a thermodynamically compatible fluid of grade three

$$\mathbf{T} = -p\,\mathbf{I} + \mu\mathbf{A}_1 + \alpha_1\mathbf{A}_2 + \alpha_2\mathbf{A}_1^2 + \beta_3(\operatorname{tr}\mathbf{A}_1^2)\mathbf{A}_1\,. \tag{3.8}$$

Let us consider simple shear flow of fluid modelled by (3.8). An easy computation yields

$$T_{xy} = \mu\kappa + 2\beta_3\kappa^3 = (\mu + 2\beta_3\kappa^2)\kappa\,, \tag{3.9}_1$$

$$T_{xx} - T_{yy} = 2\alpha_1\kappa^2\,, \tag{3.9}_2$$

$$T_{yy} - T_{zz} = (2\alpha_1 + \alpha_2)\kappa^2\,. \tag{3.9}_3$$

We see that unlike the Newtonian model, a fluid of third grade exhibits normal stress differences in simple shear flow. We also see that the fluid can shear thicken. Of course, if β_3 were not restricted to be positive due to our requirement that all flows be thermodynamically consistent, as defined here, the fluid could shear thin if $\beta_3 < 0$. However, this can lead to unpleasant features (cf. Rajagopal [18]).

We now establish the following result which is the starting point for studying the stability of the rest state of a fluid of grade three.

Theorem 9. *The motion of any fluid of grade three that is mechanically isolated for all $t \geq 0$ is such that*

$$\frac{d}{dt}\int_{\Omega_t}|\dot{\mathbf{x}}|^2\,dv + \left(\frac{\alpha_1}{2\rho}\right)\frac{d}{dt}\int_{\Omega_t}|\mathbf{A}_1|^2\,dv + \frac{\mu}{\rho}\int_{\Omega_t}|\mathbf{A}_1|^2\,dv$$

$$+ \frac{(\alpha_1 + \alpha_2)}{\rho}\int_{\Omega_t}\operatorname{tr}\mathbf{A}_1^3\,dv + \frac{\beta_3}{\rho}\int_{\Omega_t}|\mathbf{A}_1|^4\,dv = 0\,. \tag{3.10}$$

PROOF : Form the scalar product of the balance of linear momentum with the velocity and integrate over Ω_t. Use the definition of being mechanically isolated and the representation (3.8) for $\mathbf{T}$, and the result follows trivially. ∎

Now let us define a functional

$$E(t) \equiv \int_{\Omega_t} |\dot{\mathbf{x}}|^2 \, dv + \frac{\alpha_1}{2\rho} \int_{\Omega_t} |\mathbf{A}_1|^2 \, dv. \tag{3.11}$$

We see that E is zero if and only if $\mathbf{v}$ is zero in Ω_t (here we are working with smooth functions). The quantity ρE is a measure of the kinetic energy and the energy due to stretching of the fluid.

Next, we shall state without proof a lemma that is crucial to establishing results regarding the stability of flows of fluids of grade three. A proof can be found in Rajagopal [13].

Lemma. *Let* $\mathbf{A} \in \mathcal{J}_o^S$ *and let* μ *and* β *be any non–negative real numbers. If* $\alpha \in \mathbb{R}$, *then a necessary and sufficient condition that*

$$\mu |\mathbf{A}|^2 + (\alpha_1 + \alpha_2) \operatorname{tr} \mathbf{A}^3 + \beta |\mathbf{A}|^4 \geq 0, \tag{3.12}$$

is

$$-\sqrt{24 \, \mu \, \beta} \leq \alpha_1 + \alpha_2 \leq \sqrt{24 \, \mu \beta}. \tag{3.13}$$

We are now in the position to state the following corollary to Theorem 9.

Corollary. *Let us assume that a third grade fluid is mechanically isolated for all times* $t \geq 0$, *then*

$$\dot{E}(t) \leq 0 \qquad \forall \, t \in [0, \infty). \tag{3.14}$$

PROOF : Follows from the definition of mechanical isolation, the definition of $E(t)$, (3.10) and (3.13). ∎

One can obtain better results on the boundedness and stability of these fluids if the nature of the domain in which the motion is taking place and the associated boundary conditions are known. We shall first consider flows taking place within fixed rigid containers subject to the adherence condition that

$$\mathbf{v} = 0 \quad \text{on} \ \partial\Omega_t, \tag{3.15}$$

and the condition that

$$\Omega_t = \Omega \quad \forall \, t \in [0, \infty). \tag{3.16}$$

We shall call such flows cannister flows[4] in agreement with Dunn and Fosdick [19].

---

[4]If the body force field $\mathbf{b}$ is conservative, all cannister flows of an incompressible third grade fluid are mechanically isolated.

We shall investigate the possibility whether there exists a $\lambda \in \mathbb{R}^+$ such that

$$\dot{E}(t) + \lambda E(t) \leq 0 \tag{3.17}$$

for mechanically isolated cannister flows. For any $\lambda \in \mathbb{R}^+$, on using the definition (3.17) for $E(t)$, one obtains

$$\dot{E}(t) + \lambda E(t) = \frac{d}{dt} \int_{\Omega_t} |\mathbf{v}|^2 \, dv + \frac{\alpha_1}{2\rho} \frac{d}{dt} \int_{\Omega_t} |\mathbf{A}_1|^2 \, dv$$
$$+ \frac{\lambda \alpha_1}{2\rho} \int_{\Omega_t} |\mathbf{A}_1|^2 \, dv + \lambda \int_{\Omega_t} |\mathbf{v}|^2 \, dv \, .$$

On using (3.10) one obtains

$$\dot{E}(t) + \lambda E(t) = -\frac{1}{\rho} \left\{ \mu \int_{\Omega_t} |\mathbf{A}_1|^2 \, dv + (\alpha_1 + \alpha_2) \int_{\Omega_t} \operatorname{tr} \mathbf{A}_1^3 \, dv \right.$$
$$\left. + \beta_3 \int_{\Omega_t} |\mathbf{A}_1|^4 \, dv \right\} + \frac{\lambda \alpha_1}{2\rho} \int_{\Omega_t} |\mathbf{A}_1|^2 \, dv + \lambda \int_{\Omega_t} |\mathbf{v}|^2 \, dv \, . \tag{3.18}$$

Since $\mathbf{v} = 0$ on $\partial\Omega_t$ and since $\Omega = \Omega_t$, there exists a domain dependent constant C_p such that

$$\int_{\Omega_t} |\mathbf{v}|^2 \, dv \leq C_p \int_{\Omega_t} |\operatorname{grad} \mathbf{v}|^2 \, dv = \frac{C_p}{2} \int_{\Omega_t} |\mathbf{A}_1|^2 \, dv \, ; \tag{3.19}$$

the equality (3.19) being valid by virtue of divergence theorem and the fact that $\mathbf{v}$ is zero on $\partial\Omega_t$. On entering (3.19) into (3.18) we obtain

$$\dot{E}(t) + \lambda E(t) \leq -\frac{1}{\rho} \left\{ \frac{[2\mu - \lambda(\alpha_1 + \rho C_p)]}{2} \int_{\Omega_t} |\mathbf{A}_1|^2 \, dv + \right.$$
$$\left. + (\alpha_1 + \alpha_2) \int_{\Omega_t} \operatorname{tr} \mathbf{A}_1^3 \, dv + \beta_3 \int_{\Omega_t} |\mathbf{A}_1|^4 \, dv \right\} \, . \tag{3.20}$$

Let us now define the number $H(t)$ through

$$H(t) \equiv \frac{1}{\rho} \left\{ \frac{[2\mu - \lambda(\alpha_1 + \rho C_p)]}{2} |\mathbf{A}_1|^2 + (\alpha_1 + \alpha_2) \operatorname{tr} \mathbf{A}_1^3 + \beta_3 |\mathbf{A}_1|^4 \right\} \, . \tag{3.21}$$

Then, by virtue of Lemma, a necessary and sufficient condition that $H(t) \geq 0$ is that

$$\lambda(\alpha_1 + \rho C_p) \leq 2\mu \, , \tag{3.22}$$

and

$$(\alpha_1 + \alpha_2)^2 \leq 24 \frac{[2\mu - \lambda(\alpha_1 + \rho C_p)]}{2} \beta_3 \, . \tag{3.23}$$

Thus, with the choice

$$\lambda = \frac{24\mu\beta_3 - (\alpha_1 + \alpha_2)^2}{12(\alpha_1 + \rho C_p)\beta_3} \, ,$$

both (3.22) and (3.23) are met, and it is certain that $\lambda \in \mathbb{R}^+$ if

$$(\alpha_1 + \alpha_2)^2 < 24\mu\beta_3 \, . \tag{3.24}$$

Hence we have established the following result:

145

Theorem 10. *Let the cannister flow of a third grade fluid be mechanically isolated for all times* $t \geq 0$. *Further, let us suppose that the material coefficients meet the strict inequality* $|(\alpha_1 + \alpha_2)| < \sqrt{24\mu\beta_3}$. *Then there exists a* $\lambda \in \mathbb{R}^+$ *such that*

$$\dot{E}(t) + \lambda E(t) \leq 0 \qquad (3.25)$$

where

$$\lambda \equiv \frac{24\mu\beta_3 - (\alpha_1 + \alpha_2)^2}{12(\alpha_1 + \rho C_p)\beta_3} \, . \qquad (3.26)$$

The above theorem tells us that mechanically isolated cannister flows of third grade fluids whose material coefficients obey $|(\alpha_1 + \alpha_2)| < \sqrt{24\mu\beta_3}$ are asymptotically stable. Since by (3.11) we have

$$\int_{\Omega_t} |\mathbf{A}_1|^2 \, dv \leq \frac{2\rho}{\alpha_1} E(t) \, , \qquad (3.27)$$

and

$$\int_{\Omega_t} |\mathbf{v}|^2 \, dv \leq E(t) \, , \qquad (3.28)$$

then (3.25) implies that

$$0 \leq \int_{\Omega_t} |\mathbf{A}_1|^2 \, dv \leq \frac{2\rho}{\alpha_1} E(0) e^{-\lambda t} \, , \qquad (3.29)$$

$$0 \leq \int_{\Omega_t} |\mathbf{v}|^2 \, dv \leq E(0) e^{-\lambda t} \, . \qquad (3.30)$$

Next we would like to investigate the possibility that the cannister flow of a third grade fluid which is mechanically isolated for all time $t \geq 0$ is bounded from below asymptotically. This then would imply that initial disturbances cannot dissapear at any finite instant of time. We shall not provide details of the investigation but merely state the following theorem. A proof for the theorem can be found in [20].

Theorem 11. *Let the cannister flow of a third grade fluid be mechanically isolated for all time* $t \geq 0$. *Further, suppose that* Ω_t *is compact, then there exists a* $\gamma(t) \in \mathbb{R}^+$ *such that*

$$\dot{E}(t) + \gamma(t)E(t) \geq 0 \, . \qquad (3.31)$$

Thus, Theorem 10 and Theorem 11 yield the following bounds on $\int_{\Omega_t} |\mathbf{A}_1|^2 \, dv$ and $\int_{\Omega_t} |\mathbf{v}|^2 \, dv$:

$$\frac{2\rho}{\alpha_1 + \rho C_p} E(0) \, e^{-\int_0^t \gamma(\tau) \, d\tau} \leq \int_{\Omega_t} |\mathbf{A}_1|^2 \, dv \leq \frac{2\rho}{\alpha_1} E(0) e^{-\lambda t} \, , \qquad (3.32)$$

$$0 \leq \int_{\Omega_t} |\mathbf{v}|^2 \, dv \leq E(0) e^{-\lambda t} \, . \qquad (3.33)$$

All the above results are pertinent to mechanically isolated flows in cannisters. However, these results could be generalized to other situations. Details of the same can be found in [20].

Thus far, we have confined ourselves to the case

$$\mu > 0 \quad \alpha_1 > 0 \quad \beta_1 = \beta_2 = 0 \quad \beta_3 > 0, \tag{3.34$_1$}$$

$$-\sqrt{24\mu\beta_3} < \alpha_1 + \alpha_2 < \sqrt{24\mu\beta_3}. \tag{3.34$_2$}$$

Next, we shall study the temporal behavior of cannister flows in which the material coefficients meet

$$\mu > 0 \quad \alpha_1 < 0 \quad \beta_1 = \beta_2 = 0 \quad \beta_3 > 0, \tag{3.35$_1$}$$

$$-\sqrt{24\mu\beta_3} < \alpha_1 + \alpha_2 < \sqrt{24\mu\beta_3}. \tag{3.35$_2$}$$

The condition that $\alpha_1 < 0$ violates the assumption that the Helmholtz free energy has a minimum when $\mathbf{L} = \dot{\mathbf{L}} = \cdots = \overset{n-2}{\mathbf{L}} = 0$. In fact $\alpha_1 < 0$ implies that the Helmholtz free energy is a maximum when $\mathbf{L} = \dot{\mathbf{L}} = \cdots = \overset{n-2}{\mathbf{L}} = 0$, provided the process meets the Clausius–Duhem inequality.

Let us consider cannister flows that are mechanically isolated; then the power theorem reduces to

$$\frac{d}{dt} \int_{\Omega_t} |\mathbf{v}|^2 \, dv - \frac{|\alpha_1|}{2\rho} \frac{d}{dt} \int_{\Omega_t} |\mathbf{A}_1|^2 \, dv + \frac{\mu}{\rho} \int_{\Omega_t} |\mathbf{A}_1|^2 \, dv$$

$$+ \frac{(\alpha_1 + \alpha_2)}{\rho} \int_{\Omega_t} \operatorname{tr} \mathbf{A}_1^3 \, dv + \frac{\beta_3}{\rho} \int_{\Omega_t} |\mathbf{A}_1|^4 \, dv = 0, \tag{3.36}$$

where we have represented α_1 by $-|\alpha_1|$ since α_1 is negative. Let us define a function $N : [0, \infty) \mapsto \mathbb{R}$ such that

$$N(t) \equiv \frac{|\alpha_1|}{2\rho} \int_{\Omega_t} |\mathbf{A}_1|^2 \, dv - \int_{\Omega_t} |\mathbf{v}|^2 \, dv. \tag{3.37}$$

Then $N(t)$ may be interpreted roughly as a measure of the difference between the energy due to stretching and the kinetic energy of fluid. On substituting equation (3.37) into (3.36) one obtains

$$\dot{N}(t) = \frac{\mu}{\rho} \int_{\Omega_t} |\mathbf{A}_1|^2 \, dv + \frac{(\alpha_1 + \alpha_2)}{\rho} \int_{\Omega_t} \operatorname{tr} \mathbf{A}_1^3 \, dv$$

$$+ \frac{\beta_3}{\rho} \int_{\Omega_t} |\mathbf{A}_1|^4 \, dv. \tag{3.38}$$

Since μ, α_1, α_2 and β_3 meet $(3.35)_1$, by Lemma, the right side of (3.38) is non-negative and hence

$$\dot{N}(t) \geq 0.$$

Then, this implies that the difference between the average stretching of the fluid and the kinetic energy of the fluid is non-decreasing.

Next, we investigate the possibility that there exists a $\lambda \in \mathbb{R}^+$ such that

$$\dot{N}(t) - \lambda N(t) \geq 0. \tag{3.39}$$

If $N(0)$ is positive, then (3.39) implies that the difference between the average stretching of the fluid and the kinetic energy of the fluid becomes unbounded as $t \to \infty$. For any $\lambda \in \mathbb{R}^+$, (3.36) and (3.37) imply that

$$\frac{1}{\rho} \left\{ \mu \int_{\Omega_t} |\mathbf{A}_1|^2 \, dv + (\alpha_1 + \alpha_2) \int_{\Omega_t} \operatorname{tr} \mathbf{A}_1^3 \, dv + \beta_3 \int_{\Omega_t} |\mathbf{A}_1|^4 \, dv \right\}$$
$$- \frac{\lambda |\alpha_1|}{2\rho} \int_{\Omega_t} |\mathbf{A}_1|^2 \, dv + \lambda \int_{\Omega_t} |\mathbf{v}|^2 \, dv \geq 0. \tag{3.40}$$

Since $\lambda \in \mathbb{R}^+$ and $\int_{\Omega_t} |\mathbf{A}_1|^2 \, dv \geq 0$, (3.40) will be met if

$$\frac{1}{\rho} \left\{ \frac{2\mu - \lambda |\alpha_1|}{2} \int_{\Omega_t} |\mathbf{A}_1|^2 \, dv + (\alpha_1 + \alpha_2) \int_{\Omega_t} \operatorname{tr} \mathbf{A}_1^3 \, dv + \beta_3 \int_{\Omega_t} |\mathbf{A}_1|^4 \, dv \right\} \geq 0. \tag{3.41}$$

It follows from the Lemma that a necessary and sufficient condition for (3.41) to be met is that

$$(\alpha_1 + \alpha_2)^2 \leq 24 \left(\frac{2\mu - \lambda |\alpha_1|}{2} \right) \beta_3. \tag{3.42}$$

Thus, let us pick λ as

$$\lambda = \frac{24\mu\beta_3 - (\alpha_1 + \alpha_2)^2}{12|\alpha_1|\beta_3}, \tag{3.43}$$

and observe that by $(3.35)_2$ $\lambda > 0$, and that we also have

$$2\mu - \lambda |\alpha_1| \geq 0. \tag{3.44}$$

With λ thus chosen, it follows that (3.41) is met, and, moreover, so are (3.40) and (3.39). Hence, the following theorem:

Theorem 12. *Let* $\mathbf{v}$ *be a velocity field satisfying (3.36). If* $N(t)$ *is defined as in (3.37), then* $N(t)$ *is bounded below and satisfies*

$$N(t) \geq N(0) e^{\left\{ \frac{24\mu\beta_3 - (\alpha_1 + \alpha_2)^2}{12|\alpha_1|\beta_3} \right\} t}. \tag{3.45}$$

148

The above theorem tells us that in flows in which $N(0) > 0$, the difference between the average stretching and the kinetic energy, and consequently the average stretching, grows unbounded in time.

It is interesting to observe that (3.45) implies that the more viscous the fluid (i.e., the larger in μ), the more rapidly the fluid becomes unbounded in time, provided $N(0)$ is positive. Similar anomalous behavior is found in second grade fluids in which one assumes the material coefficients obey the condition

$$\mu > 0, \quad \alpha_1 < 0$$

with no condition whatsoever being imposed upon the coefficient α_2 (cf. Fosdick and Rajagopal [21])[5]. There asymptotic instability is not found to be the case, but a weaker type of growth behavior and anomalous magnification of initial data is determined.

We now state the following

Corollary. If $N(0) = 0$ and $\mathbf{v}(\cdot, 0) \neq 0$ then $N(t)$ grows unbounded in time and hence $\int_{\Omega_t} |\mathbf{A}_1|^2 \, dv$ grows unbounded in time.

PROOF : Assume $N(0) = 0$. Then, by our definition

$$\int_{\Omega_0} |\mathbf{A}_1(0)|^2 \, dv = \frac{2\rho}{|\alpha_1|} \int_{\Omega_0} |\mathbf{v}(0)|^2 \, dv \,, \tag{3.46}$$

and thus by hypothesis

$$\int_{\Omega_0} |\mathbf{A}_1(0)|^2 \, dv > 0 \,. \tag{3.47}$$

Thus $|\mathbf{A}_1(0)| \neq 0$ on a set of non–zero volume measure. Consider the function

$$g(\mathbf{A}_1) \equiv \mu |\mathbf{A}_1|^2 - \frac{|(\alpha_1 + \alpha_2)|}{\sqrt{6}} |\mathbf{A}_1|^3 + \beta_3 |\mathbf{A}_1|^4 \,. \tag{3.48}$$

Since $|\mathbf{A}_1(0)| \neq 0$ on a set of non–zero volume measure, then $g(\mathbf{A}_1(0)) > 0$ on a set of non–zero volume measure. It can also be shown that

$$\dot{N}(0) \geq \int_{\Omega_0} g(\mathbf{A}_1(0)) \, dv \,, \tag{3.49}$$

so that $\dot{N}(0) > 0$. Thus there exists a $\delta > 0$ such that $N(\bar{\delta}) > 0 \;\; \forall \bar{\delta} \in (0, \delta)$. On integrating (3.39) from $\bar{\delta}$ to t we obtain for any $t > \delta$

$$\dot{N}(t) \geq \int_{\Omega_0} N(\bar{\delta}) \, e^{\lambda(t - \bar{\delta})} \,, \tag{3.50}$$

where λ is given as in (3.43). ∎

Next we shall show that for certain cannisters $N(0)$ has to be non–negative. This then implies that in such cannisters $N(t)$ must become unbounded in time.

---

[5] That second grade fluids behave in such an anomalous fashion is the thrust of the instability results obtained in [21]. There, however, they take $\alpha_1 + \alpha_2 = 0$, as is demanded by the Clausius–Duhem inequality.

Theorem 13. *Let* $\mathbf{v}$ *be a velocity field satisfying (3.36) and let* Ω *be a rigid container whose Poincaré coefficient satisfies* $C_p(\Omega) \leq \frac{|\alpha_1|}{\rho}$. *Then* $N(0) \geq 0$ *for every field* $\mathbf{v}(\cdot, 0)$. *If* $\mathbf{v}(\cdot, 0) \neq 0$ *and* $C_p(\Omega) < \frac{|\alpha_1|}{\rho}$ *then the inequality is strict.*

PROOF : The Poincaré inequality and (3.37) imply that

$$N(t) \geq \left(\frac{|\alpha_1| - \rho C_p}{\rho} \right) \int_\Omega |\mathbf{A}_1|^2 \, dv. \tag{3.51}$$

The proof follows trivially from the above inequality. ∎

It is worth remarking that the instability result which we have established is quite non–traditional in that we show that there exists a domain where all flows are unstable. It is also easy to establish the more traditional result, that is, given a domain there exist disturbances to the rest state which grow (cf. Galdi, Padula and Rajagopal [22]).

We conclude this section with a brief discussion of uniqueness and asymptotic stability of base flows. We shall suppose for the rest of this section that (3.5) and (3.6) hold.

Let $\mathbf{v}$ be the velocity of the base flow and p be the associated pressure field, and let $(\mathbf{v}, p)$ denote a velocity–pressure pair which satisfies in Ω_t

$$\operatorname{div} \mathbf{T} + \rho \mathbf{b} = \rho \dot{\mathbf{v}}, \tag{3.52}_1$$

$$\operatorname{div} \mathbf{v} = 0, \tag{3.52}_2$$

with the velocity $\mathbf{v}$ satisfying certain boundary and initial conditions.

We wish to study the uniqueness of such a pair $(\mathbf{v}, p)$. Let $\mathbf{v}'$ be another velocity field and p' the associated pressure field such that $(3.52)_2$ is met. Further suppose that

$$\mathbf{v} = \mathbf{v}' \quad \text{on} \quad \partial \Omega_t, \tag{3.53}$$

with $\mathbf{v}'$ satisfying a certain initial condition. Let us define the velocity field $\mathbf{u}$ through

$$\mathbf{u} \equiv \mathbf{v}' - \mathbf{v}. \tag{3.54}$$

Then $\mathbf{u}$ satisfies

$$\mathbf{u} = 0 \quad \text{on} \quad \partial \Omega_t, \tag{3.55}$$

and

$$\operatorname{div} \mathbf{u} = 0 \quad \text{in} \quad \Omega_t. \tag{3.55}$$

Next, it follows from the balance of linear momentum that $(\mathbf{v}, p)$ satisfies

$$- \operatorname{grad} p + \mu \Delta \mathbf{v} + \alpha_1 \Delta \mathbf{v}_t + \alpha_1 (\Delta \boldsymbol{\omega} \times \mathbf{v}) + \alpha_1 \operatorname{grad} \left\{ (\mathbf{v} \cdot \Delta \mathbf{v}) + \frac{1}{4} |\mathbf{A}_1|^2 \right\}$$

$$+ (\alpha_1 + \alpha_2) \left\{ \operatorname{grad}(\frac{1}{4} |\mathbf{A}_1|^2) + \mathbf{A}_1 \Delta \mathbf{v} + 2 \operatorname{div}([\operatorname{grad} \mathbf{v}] \cdot [\operatorname{grad} \mathbf{v}]^T) \right\}$$

$$+ \beta_3 [\mathbf{A}_1](\operatorname{grad}(|\mathbf{A}_1|^2)) + \beta_3 |\mathbf{A}_1|^2 \Delta \mathbf{v}$$

$$+ \rho \mathbf{b} - \rho \mathbf{v}_t - \rho(\boldsymbol{\omega} \times \mathbf{v}) - \rho \operatorname{grad}(\frac{1}{2} |\mathbf{v}|^2) = 0 \,, \tag{3.56}$$

where

$$\boldsymbol{\omega} = \operatorname{curl} \mathbf{v} \tag{3.57}$$

and

$$\mathbf{A}_1 = (\operatorname{grad} \mathbf{v}) + (\operatorname{grad} \mathbf{v})^T \,. \tag{3.58}$$

We first observe that the equations of motion of a fluid of grade three, or for that matter grade two, are higher order than the Navier–Stokes equations. Thus, in general we might expect, especially when it comes to proving uniqueness theorems, that we would need to know what these additional boundary conditions might be. Interestingly, results regarding existence (cf. Galdi et. al [22]) and uniqueness (cf. Dunn and Fosdick [19], Fosdick and Rajagopal [20]) seem to need no such information! However, in finding solutions to specific problems we encounter difficulties due to paucity of boundary conditions (cf. Rajagopal [23], [24]). When considering flows in unbounded domains, it is possible to augment the usual complement of boundary conditions by requiring an asymptotic structure to the flow. The relevant issues are discussed in detail elsewhere (cf. Rajagopal et. al [25], Mansutti et. al [26], [27]).

We shall now state without proof a theorem concerning the asymptotic stability of fluids of grade three. The results regarding uniqueness follow directly from the theorem. In order to state the theorem, we shall define

$$\|f\|^2 = \int_{\Omega_t} |f|^2 \, dv \,, \tag{3.59}$$

where f represents a scalar, vector or tensor field. Next, let

$$\mathbf{U} \equiv \operatorname{grad} \mathbf{u} \,. \tag{3.60}$$

Theorem 14. *Let $(\mathbf{v}, p)$ and $(\mathbf{v}', p')$ denote the velocity pressure pairs satisfying the equations of motion (3.52) for a third grade fluid whose material coefficients meet (3.51). Further, suppose that $|(\alpha_1 + \alpha_2)| < \sqrt{24\mu\beta_3}$ and let M, $-m$ and $\mathcal{M}$ denote the maximum, minimum and absolute maximum eigenvalues of the base flow $\mathbf{v}$ on*

$\Omega \times [0, T)$, $T \leq \infty$, and let N denote the maximum eigenvalue of $\Delta \mathbf{v}$ on $\Omega \times [0, T)$. Then

$$\|\mathbf{u}(t)\|^2 + \frac{\alpha_1}{\rho} \|\mathbf{U}(t)\|^2 \leq \left(\|\mathbf{u}(0)\|^2 + \frac{\alpha_1}{\rho} \|\mathbf{U}(0)\|^2 \right) e^{\gamma t} \tag{3.61}$$

for all $t \in [0, T)$, where $\mathbf{u}$ and $\mathbf{U}$ are given as in (3.60) and γ is given by

$$\gamma = \begin{cases} \dfrac{\rho}{\rho \bar{k} + \alpha_1} D & \text{if } C \geq E, \\ \dfrac{\rho}{\alpha_1} D & \text{if } C < E, \end{cases} \tag{3.62}$$

where

$$D = \frac{2\alpha_1}{\rho}(M + m) + \bar{k} \left(\frac{\alpha_1}{\rho} N + m \right) - \frac{2}{\rho}(1 - \delta)\mu$$
$$+ \frac{6\beta_3}{\rho} \frac{(1 + 8\delta)}{4(1 - \delta)} \mathcal{M}^2 + \frac{4}{\rho} \left(|\alpha_1 + \alpha_2| \right) \mathcal{M},$$
$$C = \alpha_1 \left(\frac{\alpha_1}{\rho} N + m \right),$$
$$E = 2\alpha_1(M + m) - 2(1 - \delta)\mu + 4 \left(|\alpha_1 + \alpha_2| \right) \mathcal{M}$$
$$+ \frac{6\beta_3}{\rho} \frac{(1 + 8\delta)}{4(1 - \delta)} \mathcal{M}^2$$

and $\bar{k}$ is any upper bound for the Poincaré coefficient $C_p(\Omega_t)$ on $[0, T)$.

Corollary 1. Let $\mathcal{B}$ be a third grade fluid whose material coefficients meet the hypothesis of Theorem 14. Let $\mathbf{v}$ and $\mathbf{v}'$ be two steady flows of $\mathcal{B}$ having the same velocity boundary condition on $\partial \Omega$. Further suppose that $\mathbf{v}$ meets

$$\frac{2(1 - \delta)\mu}{\rho} > \frac{2\alpha_1}{\rho}(M + m) + \bar{k} \left(\frac{\alpha_1}{\rho} N + m \right) + 2|\alpha_1 + \alpha_2|\mathcal{M} + \frac{6\beta_3}{\rho} \frac{(1 + 8\delta)}{4(1 - \delta)} \mathcal{M}^2,$$

where

$$\delta \equiv \frac{|(\alpha_1 + \alpha_2)|}{\sqrt{24\mu\beta_3}}.$$

Then the two flows are identical.

PROOF : Since $\mathbf{v}$ and $\mathbf{v}'$ are steady, $\mathbf{u} \equiv \mathbf{v} - \mathbf{v}'$ is also steady. Hence $\|\mathbf{u}\|^2$ and $\|\mathbf{U}\|^2$ should be independent of time. However, since they are bounded from above exponentially,

$$\|\mathbf{u}(t)\|^2 = \|\mathbf{u}(0)\|^2 = 0. \tag{3.63}$$

∎

We also establish:

Corollary 2. *Let $\mathcal{B}$ be a third grade fluid whose material coefficients meet the hypothesis of Theorem 14. Let Ω be any fixed reference configuration. Then any two flows of $\mathcal{B}$ agreeing on $\partial\Omega$ for all time t and having the same velocity at $t = 0$ are identical.*

PROOF : Let us denote the two flows by $\mathbf{v}$ and $\mathbf{v}'$, and define

$$\mathbf{u} \equiv \mathbf{v} - \mathbf{v}'.$$

Then, the estimate (3.61) implies that

$$\|\mathbf{u}(t)\|^2 = \|\mathbf{U}(t)\|^2 = 0 \quad \text{for all time } t,$$

since

$$\alpha_1 \geq 0.$$

∎

4. Fluids of Rate Type and Integral Type

Fluids of Rate Type

There are many fluids of the rate type that have been proposed to model behavior of non–Newtonian fluids. Here, we shall discuss a model due to Oldroyd that contains as special cases some other interesting rate type models as well as the classical linearly viscous fluid. The Cauchy stress in a four constant Oldroyd fluid is given by (cf. Oldroyd [28])

$$\mathbf{T} = -p\,\mathbf{I} + \mathbf{S}, \tag{4.1}_1$$

$$\mathbf{S} + \Lambda_1 \left\{ \frac{\delta\mathbf{S}}{\delta t} - \frac{\alpha}{2}\left(\mathbf{A}_1\mathbf{S} + \mathbf{S}\mathbf{A}_1\right) \right\} = 2\mu\mathbf{A}_1$$

$$+ 2\mu\Lambda_2 \left\{ \frac{\delta\mathbf{A}_1}{\delta t} - \frac{\alpha}{2}\left(\mathbf{A}_1^2\right) \right\}, \tag{4.1}_2$$

where $\frac{\delta}{\delta t}$ represents the co–rotational or Jaumann time derivative defined through

$$\frac{\delta\mathbf{B}}{\delta t} = \frac{\partial\mathbf{B}}{\partial t} + [\operatorname{grad}\mathbf{B}]\mathbf{v} + \mathbf{W}\mathbf{B} - \mathbf{B}\mathbf{W}. \tag{4.2}$$

When $\alpha = 0$, the above model reduces to Jeffery's version of the Oldroyd model, while when $\alpha = 1$ it reduces to Walter's version of the Oldroyd model. When $\alpha_2 = 0$ and $\Lambda_2 = 0$, the model reduces to the classical model due to Maxwell. The model also imbeds the linearly viscous fluid model as a special case when $\alpha = 0$, $\Lambda_1 = 0$

and $\Lambda_2 = 0$. In general, $0 \leq \alpha \leq 1$. When $\alpha \neq 0$, the model allows for shear rate dependent viscosity, the presence of normal stress differences in simple shear flow and reasonable behavior in elongational flows of viscoelastic materials.

Another popular rate type model also due to Oldroyd, is usually referred to as the Oldroyd–B fluid (cf. [29]). This model includes as special cases the Maxwell model and the linearly viscous fluid model. However, the model does not allow for shear thinning or shear thickening. Recently, this model has received a lot of attention from both the theoreticians and the experimentalists in rheology.

To illustrate the complexity involved in using rate type models, we shall consider a special flow of an Oldroyd–B fluid. Existence results for the flow of an Oldroyd–B fluid have been established by Guillopè and Saut [30]. The Oldroyd–B fluid is obtained from (4.1) by setting $\alpha = 0$. In this case, the representation for the Cauchy stress $\mathbf{T}$ takes the form

$$\mathbf{T} = -p\mathbf{I} + \mathbf{S}\,,$$
$$\mathbf{S} + \Lambda_1 \left\{ \dot{\mathbf{S}} - \mathbf{LS} - \mathbf{SL}^T \right\} = \mu \left\{ \mathbf{A}_1 + \Lambda_2 \left[\dot{\mathbf{A}}_1 - \mathbf{LA}_1 - \mathbf{A}_1\mathbf{L}^T \right] \right\}\,. \qquad (4.3)$$

Let us consider the flow between two infinite parallel disks rotating about a common axis. We shall seek a rotationally symmetric solution of the form

$$\{u_r, u_\theta, u_z\} = \{rF'(z), rG(z), -2F(z)\}\,, \qquad (4.4)$$

where u_r, u_θ and u_z represent the flow in the r, θ and z directions, respectively. The flow field (4.4) automatically satisfies the constraint of incompressibility. It should be borne in mind that while the simplicity transformation (4.4) reduces the equations of motion to solving ordinary differential equations, there could be solutions to the physical problem that are not of the form (4.4). In fact, Huilgol and Rajagopal [31] and Crewther, Josza and Huilgol [32] have shown that more general solutions than those of the form (4.4) are possible.

The problem in question, as we shall see, reduces to a system of highly non–linear differential equations that admit multiple solutions. A thorough study of the problem has been provided by Ji, Rajagopal and Szeri [33] and Crewther, Josza and Huilgol [32] independently about the same time. The same problem has also been studied within the context of a lot of other non–Newtonian fluid models and the details of the same can be found in the review article by Rajagopal [34].

It follows from (4.2) and (4.3) that

$$S_{rr} + \Lambda_1 \left[r(F' S_{rr,r} - 2F'' S_{zr}) - 2F' S_{rr} - 2F S_{rr,z} + G S_{rr,\theta} \right]$$
$$= \mu \left\{ 2F' - 2\Lambda_2 \left[r^2 F''^2 - 2(F'^2 + FF'') \right] \right\},$$

$$S_{r\theta} + \Lambda_1 \left[r(F' S_{r\theta,r} - 2F'' S_{z\theta} - G' S_{zr}) - 2F' S_{r\theta} - 2F S_{r\theta,z} + G S_{r\theta,\theta} \right]$$
$$= -2\mu\Lambda_2 r^2 F'' G',$$

$$S_{rz} + \Lambda_1 \left[r(F' S_{zr,r} - F'' S_{zz}) + F' S_{zr} - 2F S_{rz,z} + G S_{rz,\theta} \right]$$
$$= \mu r \left\{ 2F'' + 2\Lambda_2 \left(3F'F'' - FF''' \right) \right\},$$

$$S_{\theta\theta} + \Lambda_1 \left[r(F' S_{\theta\theta,r} - 2G' S_{z\theta}) - 2F' S_{\theta\theta} - 2F S_{\theta\theta,z} + G S_{\theta\theta,\theta} \right]$$
$$= \mu \left\{ 2F' - 2\Lambda_2 \left[r^2 G'^2 + 2(F'^2 + FF'') \right] \right\},$$

$$S_{\theta z} + \Lambda_1 \left[r(F' S_{\theta z,r} - G' S_{zz}) + F' S_{\theta z} - 2F S_{\theta z,z} + G S_{\theta z,\theta} \right]$$
$$= \mu r \left\{ G' + 2\Lambda_2 \left(3F'G' - 2FG'' \right) \right\},$$

$$S_{zz} + \Lambda_1 \left[r F' S_{zz,r} + 4F' S_{zz} - 2F S_{zz,z} + G S_{zz,\theta} \right]$$
$$= \mu \left\{ -4F' + 8\Lambda_2 \left(FF''^2 - 2F'^2 \right) \right\}. \tag{4.5}_{1-6}$$

The balance of linear momentum reduces to

$$-\frac{\partial p}{\partial r} + \frac{\partial S_{rr}}{\partial r} + \frac{\partial S_{rz}}{\partial z} + \frac{S_{rr} - S_{\theta\theta}}{r} = \rho r \left(F'^2 - 2FF'' - G^2 \right),$$

$$\frac{\partial S_{r\theta}}{\partial r} + \frac{\partial S_{\theta z}}{\partial z} + \frac{2}{r} S_{r\theta} = 2\rho r \left(F'G - FG' \right),$$

$$-\frac{\partial p}{\partial z} + \frac{\partial S_{rz}}{\partial r} + \frac{\partial S_{zz}}{\partial z} + \frac{S_{rz}}{r} = 4\rho FF'. \tag{4.6}_{1-3}$$

The system of equations $(4.5)_{1-6}$ and $(4.6)_{1-3}$ reduces to that for a linearly viscous fluid when Λ_1 and Λ_2 are set to zero. Even in the case of the Newtonian fluid, once the equations are appropriately non–dimensionalized, multiple solutions are possible at large enough Reynolds numbers.

The system of equations $(4.5)_{1-6}$ and $(4.6)_{1-3}$ has been studied by Ji, Rajagopal and Szeri [33] and Crewther, Josza and Huilgol [32] using an analytic continuation method. There are four parameters that enter the method:

$$s = \frac{\Omega_h}{\Omega_0}, \quad \beta = \frac{\Lambda_1}{\Lambda_2}, \quad E = \frac{\nu}{\Omega_0 h^2}, \quad W = \Omega_0 \Lambda_1. \tag{4.7}$$

Here Ω_h and Ω_0 are the angular speeds of the plates at $z = 0$ and $z = h$. Then s is the ratio of the angular speeds, E is the Ekman number (inverse of the Reynolds number), β is the ratio of the relaxation time to the retardation time and W is the Weissenberg number that is the ratio of the relaxation time to a characteristic time associated with the flow.

The appropriate boundary conditions are

$$
\begin{aligned}
F(0) &= F(h) = 0 \,, \\
F'(0) &= F'(h) = 0 \,, \\
G(0) &= \Omega_0 \,, \quad G(h) = \Omega_h \,.
\end{aligned}
\tag{4.8$_{1-3}$}
$$

The system of equations is so non–linear it allows for interesting bifurcation studies. We shall not go into the details here. Between Ji, Rajagopal and Szeri [33] and Crewther, Josza and Huilgol [32] practically all the information that can be gleaned from the problem has been extracted. The Oldroyd–B fluid is but one of the many models that are used in non–Newtonian fluid mechanics. This model, however, does not allow for shear thinning or shear thickening. However, we see even in the case of this simple model (it does not have finite memory either) the mathematical problems associated with simple flows are quite challenging.

We shall conclude this chapter with a discussion of a general class of integral models that allow for finite memory effects.

Fluids of the Integral Type

Most of the integral models in use fall into the category of K–BKZ fluids (cf. Kaye [35], Bernstein, Kearsley and Zapas [36]). These models can incorporate finite memory effects. The Cauchy stress $\mathbf{T}$ in such fluids, that are incompressible, is given by

$$
\mathbf{T} = -p\mathbf{I} + 2 \int_{-\infty}^{t} \left\{ \frac{\partial U}{\partial I} \mathbf{C}_t^{-1}(\tau) - \frac{\partial U}{\partial II} \mathbf{C}_t(\tau) \right\} \, d\tau \,,
\tag{4.9$_1$}
$$

where U is the stored energy function given by

$$
U = U(I, II, t - \tau) \,,
\tag{4.9$_2$}
$$

and

$$
I = \operatorname{tr} \mathbf{C}_t(\tau) \,, \quad II = \frac{1}{2} \left[(\operatorname{tr} \mathbf{C}_t(\tau))^2 - \operatorname{tr} \mathbf{C}_t^2(\tau) \right] \,.
\tag{4.9$_3$}
$$

In general, unless the flow field is extremely simple the problems are tremendously daunting and can at best be resolved numerically. Of couse, if the stored energy function has a very simple structure then we can study more complicated flow problems. Here, we shall consider a special flow field in which the model gives rise to a simple system of equations that is amenable to analysis.

Let us consider the following flow field that is relevant to the flow in an orthogonal rheometer in which is an instrument used to measure the properties of non–Newtonian fluids (cf. Maxwell and Chartoff [37]). The instrument is essentially two parallel plates rotating with the same angular speed Ω about non–coincident axes. Due to the flow, normal stresses develop and by measuring the forces and moment on the plates we can characterize the material moduli of the fluid in question.

Let $\boldsymbol{\xi} = (\xi, \eta, \zeta)$ denote the position of a particle $\mathbf{X} = (X, Y, Z)$, at time τ. Let $\mathbf{x} = (x, y, z)$ denote the position of $\mathbf{X}$ at time t. Suppose the field is given by

$$
\begin{aligned}
\dot{x} &= v_x = -\Omega(y - g(z)), \\
\dot{y} &= v_y = \Omega(x - f(z)), \\
\dot{z} &= 0.
\end{aligned}
\tag{4.10}_{1-3}
$$

Then

$$
\begin{aligned}
\dot{\xi} &= -\Omega(\eta - g(\zeta)), \\
\dot{\eta} &= v_y = \Omega(\xi - f(\zeta)), \\
\dot{\zeta} &= 0,
\end{aligned}
\tag{4.11}_{1-3}
$$

with

$$
\xi(t) = x, \quad \eta(t) = y, \quad \text{and} \quad \zeta(t) = z.
\tag{4.12}
$$

Rajagopal [11] has shown that the motion (4.10) is a motion with constant principal relative stretch history. This immediately implies that the stress in simple fluids depends only on the kinematical tensors $\mathbf{A}_1$, $\mathbf{A}_2$ and $\mathbf{A}_3$ (cf. Wang [12]). In this flow,

$$
\mathbf{A}_3 = -\Omega^2 \mathbf{A}_1
\tag{4.13}
$$

and thus the stress depends only on $\mathbf{A}_1$ and $\mathbf{A}_2$. This simplifies the problem tremendously for though in a general flow the stress is given by the integral representation which depends on the stretch at all times τ, in this special flow, the stress only depends on the values of $\mathbf{A}_1$ and $\mathbf{A}_2$ at time t.

For the flow under consideration, a simple computation shows that

$$
\mathbf{C}_t(\tau) = 1 - \frac{\sin \Omega(t - \tau)}{\Omega} \mathbf{A}_1 + \frac{(1 - \cos \Omega(t - \tau))}{\Omega^2} \mathbf{A}_2,
\tag{4.14}
$$

$$
\begin{aligned}
\mathbf{C}_t^{-1}(\tau) = 1 &+ \frac{\sin \Omega(t - \tau)}{\Omega} \left[1 + 2(1 - \cos \Omega(t - \tau))(f'^2 + g'^2) \right] \mathbf{A}_1 \\
&- \frac{(1 - \cos \Omega(t - \tau))}{\Omega^2} \left[1 + 2(1 - \cos \Omega(t - \tau))(f'^2 + g'^2) \right] \mathbf{A}_2 \\
&+ \frac{\sin^2 \Omega(t - \tau)}{\Omega^2} \mathbf{A}_1^2 + \frac{(1 - \cos \Omega(t - \tau))}{\Omega^4} \mathbf{A}_2^2 \\
&+ \frac{\sin \Omega(t - \tau)(1 - \cos \Omega(t - \tau))}{\Omega^3} (\mathbf{A}_1 \mathbf{A}_2 + \mathbf{A}_2 \mathbf{A}_1)
\end{aligned}
\tag{4.15}
$$

157

and

$$I = II = 3 + 2[1 - \cos \Omega(t - \tau)](f'^2 + g'^2). \qquad (4.16)$$

Substituting for $\mathbf{C}_t(\tau)$, $\mathbf{C}_t^{-1}(\tau)$ and I and II in the representation (4.9), we see that depending on the specific structure of U, the equations of motion will reduce to a coupled system of ordinary differential equations. Details for special models due to Wagner and Currie have been studied in detail by Bower et al. [38] and Rajagopal et al. [39], respectively.

The appropriate boundary conditions are

$$v_x = \frac{\Omega a}{2} - \Omega y, \quad v_y = \Omega x, \quad v_z = 0 \quad \text{at } z = h,$$

$$v_x = -\frac{\Omega a}{2} - \Omega y, \quad v_y = \Omega x, \quad v_z = 0 \quad \text{at } z = 0$$

and

$$v_x \to \mp\infty, \quad v_y \to \pm\infty, \quad \text{as } x, y \to \infty \qquad (4.17)_{1-3}$$

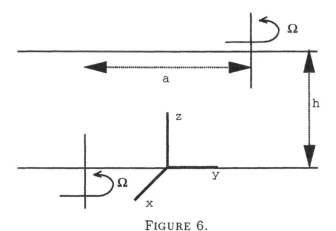

FIGURE 6.

The above boundary conditions imply that

$$f(h) = f(0) = 0,$$

$$g(h) = \frac{a}{2}, \quad g(0) = -\frac{a}{2}. \qquad (4.18)_{1,2}$$

We shall not discuss details regarding the solution to specific models but refer the reader to the review article by Rajagopal [34] on this subject. However, an interesting feature of the solution for most models is that at high Reynolds numbers the flows develop a distinct boundary layer structure adjacent to the plates, with the core essentially rotating as a rigid fluid body. This has great significance with regard to the interpretation of data from experiments in an orthogonal rheometer. Usually, it is assumed that the locus of the centers of rotation defined by $x = f(z)$ and $y = g(z)$ is a straight line joining the non–coincident axes. This is true if the flow is very slow and the Reynolds number is thus very small. However, at high Reynolds number, the locus of the centers of rotation is far from a straight line, and thus the tractions on the plate which are related to tangent to the locus of the centers of rotation at $z = 0$ and $z = h$ have a far different value than that in the case of the locus being a straight line.

If the flow field is even slightly complicated, the problems become close to intractable in the case of most K–BKZ models. For instance a great deal of effort, over a decade or more, has been expended in studying the flow of K–BKZ fluid over a slot. In this sense, flow problems involving K–BKZ fluids offer challenging opportunities to numerical analysts. With regard to more mathematical issues like existence , uniqueness and stability, practically all such questions remain unanswered for almost all models that are currently in vogue.

Bibliography

[1]   C. Truesdell and W. Noll, *The non–linear field theories of mechanics*, Handbuch der Physik, vol. III/3, Springer–Verlag, Berlin–Heidelberg–New York, 1965.

[2]   W.R. Schowalter, *Mechanics of non–Newtonian fluids*, Pergamon Press, Oxford, 1978.

[3]   R.R. Huilgol, *Continuum mechanics of viscoelastic liquids*, Hindusthan Publishing Corporation, Delhi, 1975.

[4]   C. Truesdell, *A first course in rational continuum mechanics*, vol. I, Academic Press, New York–San Francisco–London, 1977.

[5]   R.S. Rivlin and J.L. Ericksen, *Stress deformation relations for isotropic materials*, J. Rational Mech. Anal. **4** (1955), 323.

[6]   W. Noll, *A new mathematical theory of simple materials*, Arch. Rational Mech. Anal. **48** (1972), 1.

[7]   B.D. Coleman and W. Noll, *An approximation theorem for functionals with applications in continuum mechanics*, Arch. Rational Mech. Anal. **6** (1960), 355.

[8]   J.E. Dunn and K.R. Rajagopal, *Fluids of differential type: Critical review and thermodynamic analysis*, IMA Preprint Series # 1084 , University of Minnesota (1992).

[9]   W. Noll, *Motions with constant stretch history*, Arch. Rational Mech. Anal. **11** (1962), 97.

[10] R.R. Huilgol, *On the properties of motions with constant stretch history occuring in the Maxwell rheometer*, Trans. Soc. Rheology **13** (1969), 513.

[11] K.R. Rajagopal, *On the flow of a simple fluid in an orthogonal rheometer*, Arch. Rational Mech. Anal. **79** (1982), 29.

[12] C.C. Wang, *A representation theorem for the constitutive equation of a simple material in motion with constant stretch history*, Arch. Rational Mech. Anal. **20** (1965), 329.

[13] K.R. Rajagopal, *Thermodynamics and stability of non–Newtonian fluids*, Ph. D. thesis, University of Minnesota (1978).

[14] K.R. Rajagopal, *Thermodynamics and stability of non–Newtonian fluids*, Developments in Mechanics, vol. 10, Manhattan, Kansas, 1979.

[15] K.R. Rajagopal, *Boundedness and uniqueness of fluids of differential type*, Acta Ciencia Indica **18** (1982), 1.

[16] B.D. Coleman, *On the stability of equilibrium states of general fluids*, Arch. Rational Mech. Anal. **36** (1970), 1.

[17] M. Gurtin, *Modern continuum thermodynamics*, Notas De Matematica Fisica, vol. II, Universidade Federale do Rio de Janeiro, 1972.

[18] K.R. Rajagopal, *On the stability of fluids of third grade*, Archives of Mechanics **32** (1980), 867.

[19] J.E. Dunn and R.L. Fosdick, *Thermodynamics, stability and boundedness of fluids of complexity two and fluids of second grade*, Arch. Rational Mech. Anal. **56** (1982 2), 191.

[20] R.L. Fosdick and K.R. Rajagopal, *Thermodynamics and stability of fluids of grade three*, Proceedings of the Royal Society of London Ser A, 369, 1980, pp. 351.

[21] R.L. Fosdick and K.R. Rajagopal, *Anomalous features in the model of second order fluids*, Arch. Rational Mech. Anal. **70** (1978), 145.

[22] G.P. Galdi, M. Padula and K.R. Rajagopal, *On the conditional stability of the rest state of a fluid of second grade in unbounded domains*, Arch. Rational Mech. Anal. **109** (1990), 173.

[23] K.R. Rajagopal, *On the creeping flow of a second order fluid*, Journal of Non–Newtonian Fluid Mechanics **15** (1984), 239.

[24] K.R. Rajagopal, L. Tao and P. N. Kaloni, *Oscillations of a rod in a simple fluid*, J. Math. and Physical Sciences **23** (1989), 445–479.

[25] K.R. Rajagopal, A.Z. Szeri and W. Troy, *An existence theorem for the flow of a non–Newtonian fluid past an infinite porous plate*, International Journal of

Non–Linear Mechanics **21** (1986), 279–289.

[26] D. Mansutti, G. Pontrelli and K.R. Rajagopal, *Non–similar flow of a non–Newtonian fluid past a wedge*, In Press, International Journal of Engineering Science.

[27] D. Mansutti, G. Pontrelli and K.R. Rajagopal, *Flow of non–Newtonian fluids past an infinite porous plate with suction and injection*, In Press, Numerical methods in fluids.

[28] J.G. Oldroyd, *Non–Newtonian effects in steady motions of some idealized elasto–viscous liquids*, Proceedings of the Royal Society of London Series A. **245** (1958), 278.

[29] J.G. Oldroyd, *On the formulation of rheological equtions of state*, Proceedings of the Royal Society of London Series A. **200** (1950), 53.

[30] C. Guillope and J.C. Saut, *Global existence and one–dimensional non–linear stability of shearing motions of viscoelastic fluids of Oldroyd type*, Rairo–Math. Mod. and Num. Analysis – Modelisation Mathematique Et Analyse Numerique **24** (1990), 369.

[31] R.R. Huilgol and K.R. Rajagopal, *Non–axisymmetric flow of a viscoelastic fluid between rotating disks*, Journal of Non–Newtonian Fluid Mechanics **23** (1987), 423.

[32] I. Crewther, R.R. Huilgol and R. Josza, *Axisymmetric and non–axisymmetric flows of a non–Newtonian fluid between co–axial rotating disks*, Phil. Trans. R. Soc. London A **337** (1991), 467.

[33] Z. Ji, K.R. Rajagopal and A.Z. Szeri, *Multiplicity solutions in von Karman flows of Oldroyd–B fluids*, Journal of Non–Newtonian Fluid Mechanics **36** (1990), 1.

[34] K.R. Rajagopal, *Flow of viscoelastic fluids between rotating plates*, Theoretical and Computational Fluid Dynamics **3** (1992), 185.

[35] A. Kaye, College of Aeronautics, Cranfield Institute of Technology, Note No.134 (1962).

[36] B. Bernstein, E. Kearsley and L.J. Zapas, *A study of stress relaxation with finite strain*, Trans. Soc. Rheol. **7** (1963), 391.

[37] B. Maxwell and R.P. Chartoff, *Studies of a polymer melt in an orthogonal rheometer*, Trans. Soc. Rheol. **9** (1965), 51.

[38] M. Bower, A.S. Wineman and K.R. Rajagopal, *A numerical study of inertial effects of the flow of a shear thinning K–BKZ fluid in rotational flows*, Journal of Non–Newtonian Fluid Mechanics **22** (1987), 287.

[39] K.R. Rajagopal, M. Renardy, Y. Renardy and A.S. Wineman, *Flow of viscoelastic fluids between plates rotating about distinct axes*, Rheologica Acta **25** (1986), 459.

K.R. Rajagopal
Department of Mechanical Engineering
University of Pittsburgh
3700 O'Hara Street
Pittsburgh, PA 15261
U.S.A

Printed and bound by CPI Group (UK) Ltd, Croydon, CR0 4YY

23/10/2024

01778230-0015